Ultrasound Physics and its Application in Medicine

Ultrasound Physics and its Application in Medicine

ARBIN THAPALIYA
ALEC SITHOLE
MICHAEL WELSH
GASTON DANA

PALNI PRESS
INDIANAPOLIS, INDIANA

Contents

Publisher's Note

This textbook was peer-reviewed, copyedited, and published through the Private Academic Library Network of Indiana (PALNI) PALSave Textbook Creation Grants Program, which is funded by the Lilly Endowment Inc. For more information about the PALSave: PALNI Affordable Learning Program, visit the PALSave website.

If you have comments, suggestions, or corrections for this textbook, please send them to palsave@palni.edu.

PALSave

Lilly Endowment Inc.
A Private Philanthropic Foundation

Preface

A technology that began as a navigation and warfare tool in submarines during the First World War, ultrasound has grown to be a formidable, noninvasive, computerized visual tool to peer into the internal organs of the human body. Health care professionals can better interpret the images and make more accurate diagnoses by understanding the physics behind how ultrasound waves are generated, transmitted, and received. The increasing demand for the use of ultrasound technology in various biomedical fields points to a future in which medical professionals will be required to possess a general knowledge of ultrasound technology. While currently, most medical professionals rely on technicians to perform imaging studies, there can be a high degree of operator-dependent results; therefore, a solid foundation of the technological properties of ultrasound would give health care providers a competitive advantage in the future.

The discovery of ultrasound is credited to Italian physiologist Lazzaro Spallanzani, who first deduced that bats use ultrasound to navigate through echolocation. In 1938, Donald R. Griffin confirmed experimentally that bats use ultrasound for navigation. In 1826, Swiss physicist Jean-Daniel Colladon determined the speed and characteristics of sound in Lake Geneva using an underwater church bell. He successfully showed that sound waves travel faster in water than through the air. Another breakthrough came in 1880, when Pierre and Jacques Curie discovered the *piezoelectric effect*—the ability of certain materials to generate an electric charge in response to applied mechanical stress. In 1842, Austrian physicist Christian Doppler proposed that the frequency of a wave changes for an observer moving relative to its source, a process now called the *Doppler effect*. Both the piezoelectric effect and Doppler effect form the basis of the ultrasound technology used in diagnosis and therapy.

Due to the increasing need to detect icebergs and enemy submarines during the First World War, Paul Langevin developed an ultrasound transducer based on the piezoelectric effect. Karl Dussik, a neurologist and psychiatrist, started using ultrasound transducers to diagnose brain tumors in 1942.[1] Thereafter, ultrasound technology use was extended to other medical specialties as well. For example, George D. Ludwig,[2] in 1948, used it in internal medicine for diagnosing gallstones, and Donald et al.,[3] in 1962, pioneered the use of ultrasound in obstetrics and gynecology. The 1950s marked the beginning of the use of digital 2D B-mode ultrasound. Today, ultrasound technology has expanded to medical imaging in various specialties: anesthesiology, cardiology, critical care, emergency medicine, general pediatrics, internal medicine, obstetrics and gynecology, pediatric emergency medicine, physiatry, sports medicine, and surgery.

This textbook provides the theoretical and practical concepts of ultrasound and its medical applications. It focuses on sound wave concepts, transducers, imaging formation, tissue mechanics, artifacts, the Doppler effect, bioeffects, and safety. It also discusses the different uses of ultrasound for clinical and medical purposes. We believe that premedical and medical students,

1. DUSSIK KT. Ultraschall-Diagnostik, insbesondere bei Gehirnerkrankungen, mittels Hyperphonographie [Ultrasound diagnostics, especially for brain diseases, using hyperphonography]. Z Phys Ther Bader Klimanheikd. 1948 Sep–Oct;1(9–10):140–5. German. PMID: 18128879.
2. Ludwig GD, Struthers FW. Considerations underlying the use of ultrasound to detect gallstones and foreign bodies in tissue. Naval Medical Research Institute Reports; 2004 June. Project #004 001, Report No. 4.
3. Donald I. Clinical application of ultrasonic techniques in obstetrical and gynaecological diagnosis. J Obstet Gynaecol Br Emp. 1962 Dec;69:1036. PMID: 14028574.

as future clinicians, will immensely benefit from having knowledge of ultrasound physics and instrumentation in their early university life. The authors have carefully balanced theoretical concepts with practical aspects of ultrasound biomedical applications to ensure clarity on the knowledge relevant to the medical profession.

Abbreviations and Acronyms

2D	two dimensional
3D	three dimensional
4D	four dimensional
AA	abdominal aorta
AAA	abdominal aortic aneurysm
ABI	ankle brachial index
AC	acromioclavicular
AIUM	American Institute of Ultrasound in Medicine
AK	above the knee
ALARA	As Low As Reasonably Achievable
A-Mode	amplitude mode
AR	anterior recess
ARDMS	American Registry for Diagnostic Medical Sonography
ARDS	acute respiratory distress syndrome
ATA	anterior tibial artery
ATV	anterior tibial vein
AUG	augmentation
BA	brachial artery
BART	blue away, red toward
BHCG	beta human chorionic gonadotropin
BI	biceps
BK	below the knee
B-Mode	brightness mode
BPP	biophysical profile
BR	brachialis
BUN	blood urea nitrogen
CBC	complete blood count
CCA	common carotid artery

CET	common extensor tendon
CFA	common femoral artery
CFV	common femoral vein
COMP	compression
COVID-19	Coronavirus disease 2019
CPR	cardiopulmonary resuscitation
CSF	cerebrospinal fluid
CT	computerized tomography
CW	continuous wave
DPA	dorsalis pedis artery
ECA	external carotid artery
EDC	estimated date of confinement
EDVol	end-diastolic volume
EF	ejection fraction
ESVol	end-systolic volume
FAST	Focused Assessment with Sonography in Trauma
FO	foramen ovale
FV	femoral vein
GASTROC V	gastrocnemius vein
GSV	great saphenous vein
ICA	internal carotid artery
ICU	intensive care unit
IUGR	Intrauterine Growth Retardation
IVP	intravenous pyelogram
LA	left atrium
LCA	left coronary artery
LV	left ventricle
MI	mechanical index
M-Mode	motion mode

MN	median nerve
MRI	magnetic resonance imaging
MSK	musculoskeletal
PAD	peripheral artery disease
Pero A	peroneal artery
PERO V	peroneal vein
PFO	patent foramen ovale
pH	potential or power of hydrogen
PI	pulsatility index
Pop A	popliteal artery
POP V	popliteal vein
PPE	personal protective equipment
Prof A	profunda femoral artery
PROF V	profunda femoral vein
Prox	proximal
PSV	peak systolic velocity
PTA	posterior tibial artery
PTV	posterior tibial vein
PW	pulsed wave
RA	right atrium
RN	radial nerve
RV	right ventricle
SAG	sagittal
SARS-CoV-2	Severe Acute Respiratory Syndrome Coronavirus 2
SFA	superficial femoral artery
SFJ	saphenofemoral junction
SMA	superior mesenteric artery
SPI	Sonography Principles and Instrumentation

SSV	small saphenous vein
STIC	Spatio-Temporal Imaging Correlation
SVol	stroke volume
TCD	transcranial Doppler
TI	thermal index
TIB	thermal index for bone
TIC	thermal index for cranial bone
TIS	thermal index for soft tissue
Trans	transverse
TRV	transverse
UCL	ulnar collateral ligament
U.S. FDA	United States Food and Drug Administration
Vert	vertebral artery
VR	velocity ratio

List of Definitions

1. **absorption**—The attenuation process involving ultrasound energy lost to tissue in the form of heat.
2. **acoustic**—Having to do with sound.
3. **amniotic fluid**—A liquid that surrounds the fetus in the uterus.
4. **amplifier**—An electronic device that enhances a signal, making the output signal stronger than the input signal.
5. **amplitude** (of a wave)—Maximum displacement of a point on a wave or vibrating object relative to equilibrium position.
6. **amplitude mode**—A one-dimensional display mode used in diagnostic sonography in which the echo depth is on the horizontal axis and the amplitude is on the vertical axis.
7. **anechoic**—Areas that appear dark on ultrasound because they do not reflect sound waves back to the transducer.
8. **anesthesiology**—A branch of medical science dealing with anesthesia and anesthetics.
9. **aneurysm**—An unexpected balloon-like bulge in an artery. If it bursts or leaks, it can cause dangerous bleeding or even death. According to the National Institutes of Health, about 13,000 people die yearly in the United States from abdominal aortic aneurysms.
10. **angle of insonation**—Within sonography, the angle that the ultrasound beam makes relative to the tissue or organ of interest.
11. **anisotropy**—An artifact on ultrasound images caused by variations in the angles of insonation produced by varying properties of tissue.
12. **anterior**—Relating to the front (for example, the kneecap is located on the anterior side of the leg).
13. **artifact**—A false portrayal of image anatomy usually characterized by an inappropriate brightness, shape, size, or position

14. “**As Low As Reasonably Achievable**”—Refers to responsibly exposing the patient to the minimum energy possible to obtain appropriate diagnostic information.
15. **atherosclerosis**—A disease in which plaque (buildup of fat, cholesterol, calcium, or other products) builds up, subsequently narrowing the arteries and thus limiting the flow of oxygenated blood to organs and other parts of the body. This condition can lead to heart attack, stroke, or even death.
16. **attenuation**—The decrease in the energy and amplitude of a wave as it passes through a medium due to the effects of absorption, reflection, and scattering. As the beam passes through tissue, its intensity decreases over time.
17. **axial resolution**—The minimum distance between two reflectors along the sound path that can be resolved to produce separate sound waves.
18. **Baker’s cyst**—A fluid-filled growth behind the knee.
19. **bioeffects of ultrasound**—Within sonography, refers to the adverse effects of ultrasound beams on tissues (including soft tissue, body fluids, and bone).
20. **biohazard**—Biological organisms or substances that can threaten human health.
21. **biopsy**—A procedure medical professionals use to remove a piece of tissue or a sample of cells from the body for laboratory analysis.
22. **blood pressure**—Pressure created by blood against the walls of the arteries as the heart pumps it through the body.
23. **brightness mode**—A two-dimensional display mode used in diagnostic sonography in which the image is composed of bright dots representing the ultrasound echoes. The brightness of each dot is determined by the amplitude of the returned echo signal.
24. **calcification**—A process of calcium buildup in body tissue that causes the tissue to harden.
25. **cardiac arrest**—An abrupt inability of the heart to contract and generate a pulse and adequate blood flow leading to starvation

of blood in the brain and other organs.

26. **cardiology**—A medical specialty that involves finding, treating, and preventing diseases of the heart and blood vessels.
27. **cardiopulmonary resuscitation**—A lifesaving procedure performed by combining rescue breathing (mouth-to-mouth) and chest compressions to temporarily pump enough blood to the brain when the heart stops beating and until specialized treatment is available.
28. **caudal**—*See inferior.*
29. **cavitation**—Production and dynamics of bubbles caused by sound waves. The bubbles oscillate inside the tissue due to the rapidly changing acoustic pressure.
30. **circle of Willis**—A complete ring of arteries at the base of the brain that is formed by the cerebral and communicating arteries.
31. **color Doppler display**—Two-dimensional, real-time Doppler shift information superimposed on a real-time, grayscale, anatomic cross-sectional image. Flow directions display different colors.
32. **comet tail**—A series of closely spaced reverberation echoes.
33. **compression**—A region in a longitudinal wave (like sound waves) where the particles are closest together.
34. **computerized tomography**—An imaging technique that combines a series of X-ray images taken from different angles around the body.
35. **continuous wave Doppler**—An imaging mode in which the transducer uses two crystals: one for continuous transmission and the other for reception of ultrasound signals. This technique allows measurements of high-velocity flows, especially in the heart.
36. **contrast resolution**—The ability to distinguish echoes of slightly different intensities on a grayscale display.
37. **Coronavirus disease 2019**—An infectious disease caused by the Severe Acute Respiratory Syndrome Coronavirus 2 (SARS-CoV-2) virus that was discovered in 2019.

38. **coupling medium**—Within sonography, a gel that is applied between the tissue surface, typically skin, and the ultrasound transducer to displace air at the boundary surface and transmit sound waves through the skin.
39. **cranial**—*See superior.*
40. **critical care**—A branch of medicine focused on the care of people with life-threatening or critical injuries and illnesses requiring more specialized care.
41. **cross section**—A transverse cut through a structure or tissue.
42. **cycle**—Within sonography, represents the combination of one rarefaction and one compression of a sound wave.
43. **cyst**—An abnormal sac or cavity in the body containing fluid or semisolid material.
44. **deep**—Position that is closer to the interior center of the body.
45. **dermatology**—A branch of medicine dealing with the skin, its structure, and its diseases.
46. **diastole**—A phase in the cardiac cycle during which the heart relaxes and allows blood to refill.
47. **diffraction**—Refers to the bending of an ultrasound beam when it passes around an edge or through a narrow opening.
48. **distal**—Away from or farthest from the trunk or the point or origin of a part (for example, the hand is located at the distal end of the forearm).
49. **Doppler angle**—The angle between the sound beam and the flow direction.
50. **Doppler effect**—A shift in frequency received by the transducer secondary to the sound wave reflecting off a moving object, such as a moving collection of red blood cells. Within sonography, this phenomenon is used to measure the flow velocity of the blood during echocardiography.
51. **Doppler equation**—Relates the various parameters that affect the Doppler shift frequency. The equation is

$$f_D = \frac{v \cos \theta}{c} f_S,$$

where f_S is the Doppler shift frequency, v is the receiver (target) velocity, θ is the Doppler angle, and c is the speed of sound in tissue, which is 1540 m/s.

1. **Doppler shift frequency**—The amount of change in frequency caused by the Doppler effect.
2. **dorsal**—*See posterior.*
3. **duplex imaging**—A process of combining two-dimensional real-time imaging with Doppler signal processing.
4. **echo**—Within sonography, a reflected ultrasound signal. The signals are converted by the processor into gray dots on the ultrasound display.
5. **echocardiography**—A diagnostic ultrasound test used to evaluate cardiac anatomy and physiology.
6. **eddies**—Regions of circular flow patterns present within turbulent flow patterns.
7. **edema**—Swelling caused by too much fluid trapped in the body's tissues.
8. **effusion**—An excessive accumulation of fluid in a region or a structure.
9. **emergency medicine**—A medical specialty involving care for patients with acute illnesses or injuries requiring immediate medical attention.
10. **emphysema**—A lung condition that causes shortness of breath due to damaged air sacs in the lungs.
11. **fascia**—A thin casing of connective tissue that surrounds and holds every organ, blood vessel, bone, nerve fiber, and muscle in place.
12. **fibrosis**—Scarring and hardening of tissues and organs.

13. **flexion**—Bending of an arm or leg. It's a physical position that decreases the angle between the bones of the limb at a joint.
14. **focal length**—The distance from the scanning face of a transducer to the focus.
15. **focus**—A narrow point in the cross section of the ultrasound beam. To improve the resolution at a given depth, the transducer must be focused in order to narrow the beam width. This is performed by varying the number of transmitting and receiving elements and delaying the signal once received.
16. **frame**—A single image produced by one complete scan of the sound beam.
17. **frame rate**—The number of frames of echo information stored each second.
18. **frequency**—The number of cycles per second. In diagnostic ultrasound, it is usually demonstrated in megahertz.
19. **friction**—Within sonography, the force that resists the relative motion between blood and the cardiovascular walls.
20. **gain**—The degree of amplification of the returning echo.
21. **gate**—A device that allows only echoes from selected depths to pass.
22. **gestation**—The process or period of development inside the womb between conception and birth.
23. **grating**—The process in which an input ultrasound beam splits into multiple beams after passing tissue, resulting in the formation of side lobe artifacts.
24. **grayscale**—The range of brightness between white and black. The dark color indicates the total absence of reflected light, and the bright represents reflected light.
25. **gynecology**—A specialty of the female reproductive system, including the diagnosis and treatment of conditions, disorders, and diseases.
26. **harmonics**—Frequencies that are even and odd multiples of one another.
27. **hertz**—Unit of frequency, one cycle per second.
28. **hydrocephalus**—A neurological disorder caused by an

abnormal buildup of cerebrospinal fluid in the ventricles or cavities deep within the brain.

29. **hydronephrosis**—Swelling of a kidney due to urine failing to drain properly from the kidney to the bladder.
30. **hyperechoic**—Areas on the image with more reflected echoes that appear brighter than the surrounding tissue.
31. **hypertension**—High systolic and/or diastolic blood pressure.
32. **hypoechoic**—Areas on the image with fewer reflected echoes that appear darker than the surrounding tissue.
33. **hysterectomy**—A surgical removal of the uterus.
34. **impedance**—The resistance to the propagation of ultrasound waves through tissues. It is determined by the density of the tissue and the sound propagation speed.
35. **inertia**—Resistance to change in the state of either rest or motion or the tendency to keep moving in a straight line at a constant velocity.
36. **inferior**—Away from the head; lower (for example, the foot is part of the inferior extremity).
37. **intensity**—The power per unit area, usually expressed in milliwatts per square centimeter in sonographic discussions. It is the ultrasound power transferred per unit area.
38. **internal medicine**—A branch of medicine that deals with long-term care involving common and complex illnesses such as the diagnosis and treatment of cancer, infections, and diseases affecting the heart, blood, kidneys, joints, and digestive, respiratory, and vascular systems in adolescents, adults, and elderly individuals.
39. **intima-media thickness analysis**—An ultrasound measurement of the thickness of the two inner layers of the carotid arteries—the intima and media. Abnormal thickening of the arterial walls of the carotid arteries indicates a high risk for heart attack or stroke.
40. **intrarenal perfusion**—Fluid injection into the renal blood vessels to or from the kidneys.
41. **laminar flow**—Demonstration of fluid moving in an organized

direction as if it were made up of layers that glide over one another.

42. **lateral**—Away from the midline of the body (for example, the little toe is located at the lateral side of the foot).
43. **lateral resolution**—Refers to reflectors that lie perpendicular to the axis of the ultrasound beam. The resolution is related to the beam width; that is, the wider the beam, the poorer the lateral resolution. An ultrasound beam is the narrowest at its focal length, and this is where the lateral resolution will be the best.
44. **lateral rotation**—Lateral rotation of the upper limb at the shoulder or lower limb at the hip involves turning the anterior surface of the limb away from the midline of the body. Lateral rotation is also referred to as exterior rotation.
45. **linear array**—An array made of rectangular elements arranged in a line.
46. **linear phased array**—A linear array that operates by applying voltage pulses to all elements with small time differences to direct sound waves out in various directions.
47. **linear sequenced array**—A linear array that operates by applying voltage pulses to groups of elements in a sequence.
48. **Lister's tubercle**—A bony prominence located at the distal end of the radius.
49. **lumen**—An inner open space or cavity of a tubular structure, as of a blood vessel or an intestine.
50. **maternity**—The period during and immediately after pregnancy.
51. **mechanical index**—A value used to estimate the cavitation effect during exposure to ultrasound energy.
52. **medial**—Relating to the midline of the body (for example, the middle toe is located at the medial side of the foot).
53. **medial rotation**—Medial rotation of the upper limb at the shoulder or lower limb at the hip involves turning the anterior surface of the limb toward the midline of the body. Medial rotation is also referred to as internal rotation.

54. **mirroring**—An artifact on a spectral display caused by an inability to separate the forward and reverse signal processing channels properly.
55. **morphology**—The scientific study of the structure and form of animals and plants.
56. **Morrison's pouch**—Area between the liver and right kidney.
57. **motion mode**—A one-dimensional position-versus-time ultrasound view of all changing or moving reflectors. This is used in echocardiography.
58. **obstetrics**—The branch of medicine that deals with caring for and treating individuals during pregnancy, childbirth, and the postbirth period as well as the care of the newborn.
59. **oligohydramnios**—A condition where there is reduced amniotic fluid around the baby during pregnancy.
60. **ophthalmology**—A branch of medicine involving anatomy, physiology, and conditions of the eye.
61. **paracolic gutters**—Areas between the colon and the abdominal wall.
62. **pediatrics**—A specialty providing medical care to infants, children, and adolescents.
63. **perinatologist**—An obstetrician-gynecologist who specializes in high-risk pregnancies.
64. **period**—Time per cycle.
65. **physiatry**—A branch of medicine that deals with the treatment of injuries or illnesses of the nerves, muscles, and bones. Also known as physical medicine and rehabilitation.
66. **piezoelectric effect**—The ability of certain crystals to generate an electric charge in response to applied mechanical stress. Piezoelectric materials produce electrical signals when deformed or vice versa (i.e., the crystal changes shape and vibrates when a voltage is applied to it). These crystals are the primary components of ultrasound transducers.
67. **pixel**—Picture elements; the unit into which imaging information is divided for storage and display.
68. **polyhydramnios**—A condition when there is too much

amniotic fluid around the baby during pregnancy.

69. **posterior**—Relating to the back (for example, the shoulder blades are located on the posterior side of the body).
70. **potential or power of hydrogen**—A measure of hydrogen ion concentration representing the acidity or basicity of aqueous or other liquid solutions.
71. **pouch of Douglas**—The caudal extension of the peritoneal cavity. It is the rectovaginal pouch in the female and the rectovesical pouch in the male.
72. **propagation speed**—The speed at which a wave travels through a medium and is equal to the product of the wavelength and frequency. The average propagation speed of sound waves in a soft tissue is 1540 meters/second.
73. **proximal**—Toward or nearest the trunk or the point of origin of a part (for example, the proximal end of the femur joins with the pelvic bone).
74. **pulsatility index**—A dimensionless parameter that is used to assess pulsatility and is defined as the difference between maximum and minimum blood flow velocity, normalized to the average velocity.
75. **pulsed Doppler**—A Doppler mechanism that transmits a coherent burst of sound waves into the tissue and waits for returning echoes from a specific location, then processes these echoes for Doppler shift frequency content.
76. **pulse repetition frequency**—The number of pulses per second.
77. **pulse repetition period**—Interval of time from the beginning of one pulse to the beginning of the next.
78. **rarefaction**—A region in a longitudinal wave (like sound waves) where the particles are farthest apart.
79. **real-time display**—Immediate display of an ultrasound scan of moving structures made possible by the use of higher frame rates.
80. **reflection**—Portion of the sound returned from a medial boundary (i.e., refers to the bending of the ultrasound beam at the boundary surface between two mediums with different

acoustic impedances).

81. **refraction**—Change of sound direction on passing from one medium to another.
82. **regurgitation**—The flow of a fluid through a vessel or valve in the body in a direction opposite to normal.
83. **renal arterial resistive index**—An index used to assess the risk for renal arterial disease. It is commonly used to indicate renal function, pathology, prognosis, and responsiveness to steroid therapy in chronic kidney disease patients.
84. **resolution**—Within sonography, represents the ability of an ultrasound system to distinguish between two closely spaced reflectors as separate structures.
85. **reverberation**—Artifact resulting from false echoes produced due to repeated reflections between two interfaces with a high acoustic impedance mismatch.
86. **sagittal plane or view**—An anatomical plane that divides the body into right and left sections.
87. **scattering**—Spreading of ultrasound beam in different directions as it strikes a boundary or interface between two small structures.
88. **shadowing**—Reduction in echo amplitude from reflectors behind a strongly reflecting or attenuating structure.
89. **sonography**—The process of visualizing body structures using ultrasound.
90. **spatial**—Relating to the position, area, and size of things.
91. **spatial resolution**—The minimum distance between two adjacent features that can be detected by the imaging system.
92. **spatiotemporal**—Relating to both space and time.
93. **spectral analysis**—Separation of frequencies in a Doppler signal for display as a Doppler spectrum in the form of a neat graph.
94. **specular reflection**—An optical process involving reflections from a large, smooth boundary.
95. **stenosis**—An abnormal narrowing of a blood vessel or other tubular organ or structure such as foramina and canals.

96. **superior**—Toward the head end of the body; upper (for example, the hand is part of the superior extremity).
97. **supine**—A position in which a person lies on their back with their face and torso pointing up.
98. **surgery**—An invasive use of instruments to test, remove, or adjust an individual's original anatomy for treatment of illness, injury, or disease or cosmetic purposes.
99. **systemic hemodynamics**—Study of the relationship between arterial blood pressure and blood flow. Systemic hemodynamic state is the mean value of blood pressure and the mean value of blood flow over one heartbeat interval.
100. **systole**—Phase of the cardiac cycle in which the heart contracts and pumps out the blood.
101. **temporal**—Relating to time.
102. **thermal index**—A value used as an indicator for the potential risk of tissue heating due to exposure to ultrasound.
103. **thrombosis**—The development of blood clots inside blood vessels that obstruct blood flow in the circulatory system.
104. **thrombus**—A blood clot in the circulatory system.
105. **transducer**—A device that converts variations in acoustic pressure to an electrical signal or vice versa.
106. **transmitter**—A device that uses an electrical current to generate acoustic vibrations or pulses.
107. **transverse plane or view**—An anatomical plane that divides the body into superior and inferior sections.
108. **trimester**—A period of three months during gestation. The human gestation period has three trimesters (first, second, and third).
109. **turbulent flow**—A nonlaminar flow condition in which the kinetic energy of flow creates vortices. The overall fluid motion is complex and chaotic.
110. **urology**—The study or medical treatment of diseases and conditions of the urinary tract system and certain reproductive organs.
111. **valves**—A flap or leaflet, usually inside a tubular structure, that

acts as a one-way inlet and permits fluids or gases to flow in just one direction.

112. **ventilator**—A device that supports or re-creates the process of breathing by pumping air into the lungs when a person is not able to breathe enough on their own.
113. **ventral**—*See anterior.*
114. **viscosity**—A measure of the resistance of a fluid to flow. It is an indicator of the internal friction of a moving fluid. For example, a fluid with high viscosity flows slower than one with low viscosity.

1. Basic Principles of Ultrasound

1.1 Learning Objectives

After reviewing this chapter, you should be able to do the following:

1. Define ultrasound and describe its characteristics as a form of energy.
2. Explain the principles of sound wave propagation, including frequency, wavelength, amplitude, and velocity.
3. Describe the piezoelectric effect and how it is used in ultrasound transducers.
4. Explain the difference between longitudinal and transverse waves and how they relate to ultrasound.
5. Understand the concept of acoustic impedance and its effect on ultrasound imaging.
6. Describe the interaction of ultrasound with tissue, including absorption, reflection, and scattering.

1.2 Introduction

This chapter introduces the fundamental principles of sound wave propagation, including frequency, wavelength, amplitude, and velocity and how these relate to ultrasound imaging. The chapter also covers the interaction of ultrasound with tissue, including absorption, reflection, and scattering.

1.3 Why Study Ultrasound?

Ultrasound has a wide range of medical applications. These include its use in obstetrics (monitoring the progress of pregnancy), oncology (monitoring the growth of tumors), cardiology (visualizing heart function and physiology), and biopsy (guiding needles in various procedures) and as a rehabilitation modality. For example, today, an estimated 60–70% of pregnant women in the United States undergo ultrasound examinations during pregnancy. An estimated 250 million fetal ultrasound examinations are performed annually in the United States.[1] The examination aims to assess fetal abnormalities, confirm the site of pregnancy within the uterus, and determine gestational age.

Due to the increasing use of ultrasound in health care, most medical schools are transforming curricula to include medical ultrasound applications. In addition, most medical professions require successful completion of a Sonography Principles and Instrumentation (SPI) examination and some specialty examinations administered by the American Registry for Diagnostic Medical Sonography (ARDMS). The SPI examination requires a sound knowledge of the physical principles of ultrasound and imaging,

1. Nelson TR, Fowlkes JB, Abramowicz JS, Church CC. Ultrasound biosafety considerations for the practicing sonographer and sonologist. J Ultrasound Med. 2009 Feb;28(2):139–50. doi: 10.7863/jum.2009.28.2.139. PMID: 19168764.

which includes an understanding of the physics of ultrasound and imaging techniques. This requirement is appalling to most clinicians with no physics background. On the other hand, depending on the individual's interest, a specialty examination will be given in one or more of the following areas: abdomen, breast, echocardiography, obstetrics, gynecology, pediatric, vascular technology, or musculoskeletal ultrasound. Most hospitals recognize ARDMS credentials as requirements for clinical sonographers and physicians.

Ultrasound images in medical imaging are generated from sound waves reflected from different tissues and organs and converted into electrical signals, which a computer processes to create an image displayed on a screen. This technology helps health care professionals visualize internal structures and diagnose medical conditions. To make it easier to understand the operation of a medical ultrasound machine, we will first discuss some basic physics principles.

1.4 Mechanical Vibrations and Waves

A mechanical vibration is a back-and-forth motion. When vibrations affect the media around them, waves are generated. These waves transport energy from one point to another. If a single vibratory disturbance moves from one point to the other, it is called a *pulse*. A back-and-forth motion that occurs repeatedly is called a *periodic motion*.

A mechanical wave requires a material medium (such as a solid, liquid, or gas) to propagate through; its speed depends on the properties of that medium. Mechanical waves fall into two classes: longitudinal and transverse waves. For a transverse wave, the displacement of the medium is perpendicular to the direction of the motion of the wave. In a longitudinal wave, the displacement of the medium is in the same direction as wave motion. One example of a longitudinal wave is sound. These waves are similar to the motion of a pulse on a slinky, as illustrated in Figure 1-1.

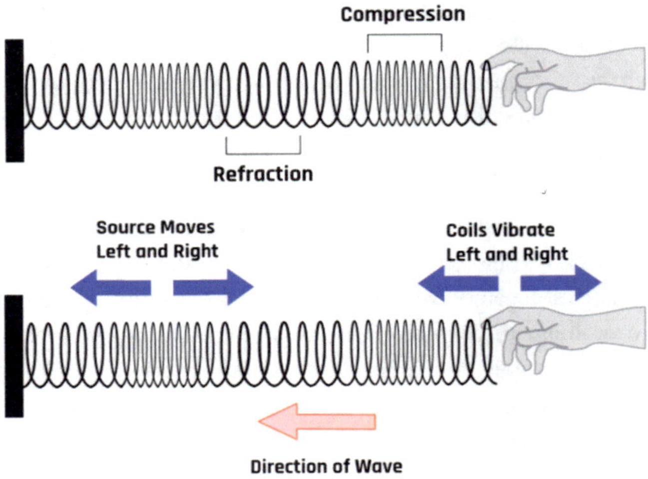

Figure 1-1: Characteristics of a longitudinal wave on a slinky.

1.5 Characteristics of Sound Waves

A sound wave comprises alternating regions of low and high pressures. The waveform is a sinusoidal wave function in which the crests and troughs represent high- and low-pressure regions, respectively, as shown in Figure 1-2.

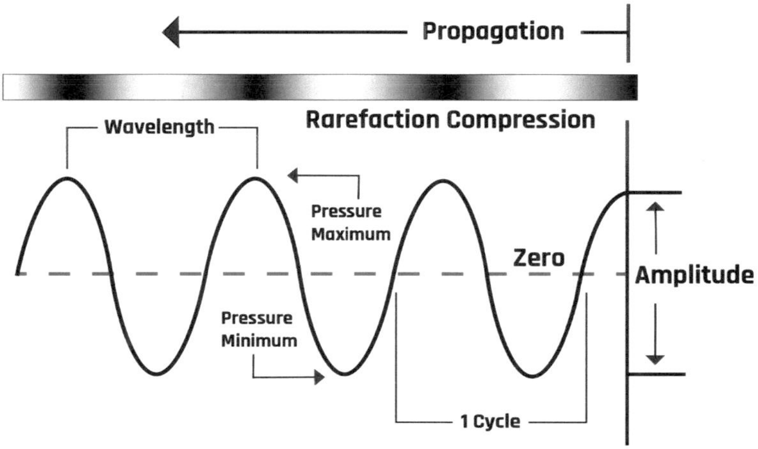

Figure 1-2: Characteristics of a longitudinal wave.

1.5.1 Amplitude and Wavelength

The maximum displacement or height from the horizontal axis, the equilibrium position, is the amplitude (A) of the wave. The distance between two successive points in the same phase is the wavelength (λ). For example, the wavelength is the distance between neighboring peaks, neighboring troughs, or any two points where the wave returns to the same shape, as shown in Figure 1-3.

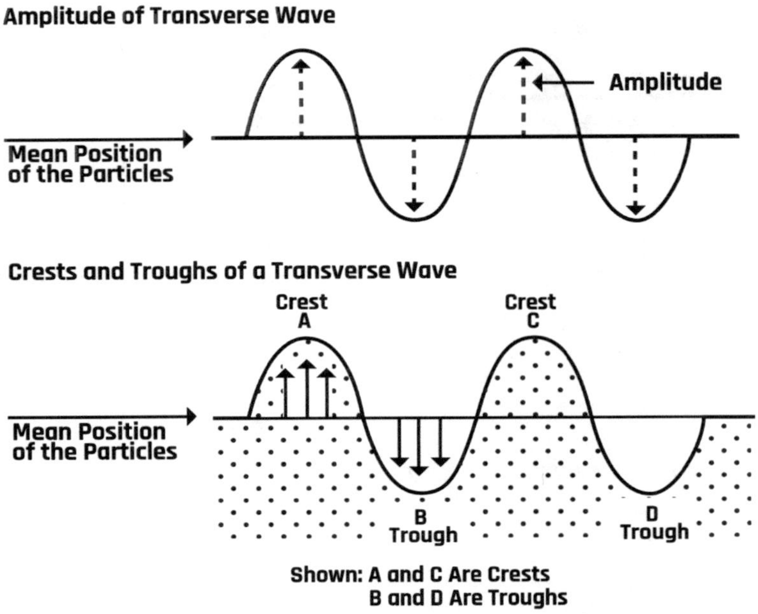

Figure 1-3: Transverse wave characteristics.

1.5.2 Period and Frequency

The time for one complete cycle is the period (T); it is related to the number of oscillations (complete cycles) per unit time, called the frequency (f), as follows: $T = 1/f$. The unit of frequency is the cycle/second = 1/s or s^{-1}. One cycle per second = 1 hertz (Hz).

1.5.3 Wave Velocity

The rate at which the waveform changes position with respect to time is called the velocity (v) of the wave. In mathematical form, v = f λ. The wave velocity depends on the characteristics of the medium in which it travels. The velocity has a unit of meters per second if the frequency (f) is in hertz and the wavelength (λ) is in meters. The speed of sound is fastest in solids, slower in liquids, and slowest in air. Sound does not travel through free space or vacuums due to the lack of a medium.

1.6 Audible and Ultrasound Waves

Sound waves are classified into three categories based on their frequency: *infrasonic*, *audible*, and *ultrasonic waves*.

Infrasonic waves: These are waves with frequencies of less than 20 Hz; they are not audible to the human ear. Various natural and man-made sources produce them, including earthquakes, volcanoes, thunderstorms, and industrial machinery. Animals such as elephants, whales, and alligators also produce these waves to communicate over long distances.

Audible waves: These are sound waves with frequencies that the human ear can hear. The audible frequencies fall in the range of about 20 Hz to 20 kHz. The audible range of sound can vary between individuals due to various factors. Age is one factor that can affect a person's ability to hear some frequencies, as the ear's sensitivity decreases with age. For example, younger people are likely to hear sounds of up to 20 kHz, while older people mostly hear frequencies of far less than 20 kHz. Certain diseases and medical conditions, such as otitis media, otosclerosis, or Meniere's disease, can also affect a person's hearing ability. Additionally, individuals with hearing impairments or disabilities may have a reduced hearing range due to damage to the ear or nerve pathways involved in hearing.

Ultrasonic waves (also called *ultrasound*): These are waves with a frequency of more than 20 kHz. Humans cannot hear ultrasound, but bats use these waves for navigation. However, medical professionals use these waves to examine or image the different parts of the human body, a practice also known as *sonography*. Medical ultrasound imaging typically uses the 3.5 to 20 MHz frequency range.

1.7 Quantifying Ultrasound

The amount of sound energy flux per unit of time is called the *sound intensity* (I). For a point source generating sound with *acoustic power* (P), the intensity (I) at distance (r) from the source obeys the inverse square law:

$$I = \frac{P}{4\pi r^2}.$$

The *acoustic power* of an ultrasound wave is the quantity of energy generated per unit of time. The standard unit of acoustic power is the watt (W), and 1 watt = 1 joule per second. Therefore, the unit of sound intensity is W/m^2. The intensity equation shows that sound intensity decreases as the square of the distance from the point source. We all know that sound loudness (ear perception of sound intensity level) decreases as we move away from the source, as illustrated in Figure 1-4.

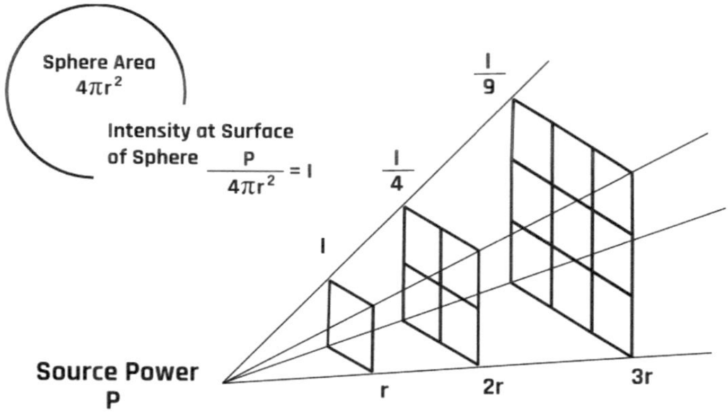

Figure 1-4: The intensity of a wave decreases inversely with the square of the distance from its source.

The *sound intensity level* (also called the *sound acoustic level*) is commonly measured relative to the *standard threshold of hearing intensity* (I_0) in decibels. A decibel is a dimensionless quantity (no units) represented as dB, which is based on the logarithmic scale. In mathematical form, the sound intensity level (ß) is expressed as

$$ß(\mathrm{dB}) = 10 \log \frac{I}{I_0},$$

where $I_0 = 10^{-12}$ W/m^2, which is the faintest audible sound intensity.

1.8 Propagation of Ultrasound Through Tissues

Clinicians examine body tissue structures using ultrasound waves with 2 to 20 MHz frequencies. Inside the tissue structures, the waves propagate through the medium by vibrating molecules of the medium. In soft tissues, the propagation velocity is relatively constant at 1540 m/s. This velocity value is used by ultrasound machines for all human tissue. Ultrasound waves propagating through tissue undergo reflection, refraction, attenuation, scattering, and diffraction.

1.8.1 Reflection

Like any wave, ultrasound waves are reflected at tissue boundaries and interfaces. The transducer detects these reflected waves, and piezoelectric signals are generated and processed into an image form via a computerized processing unit. These signals form the basis of all ultrasound imaging. The number of reflected waves detected by the transducer depends on the angle of incidence at the band boundary and the difference in acoustic impedance between the two tissues traversed by the beam. More details about the acoustic impedance will be discussed later. However, it represents the resistance of a tissue to the passage of ultrasound. Typically, a propagating ultrasound wave is split into two components, as Figure 1-5 illustrates.

If the wave traverses from medium 1 (with acoustic impedance Z_1) to medium 2 (with acoustic impedance Z_2), the reflection coefficient is

$$\frac{P_r}{P_i} = \frac{Z_2 \cos \theta_i - Z_1 \cos \theta_t}{Z_2 \cos \theta_i + Z_1 \cos \theta_t},$$

where θ_i is the angle of incidence, θ_r is the reflected angle, and θ_t is the angle of transmission. P_r and P_i represent the reflection and incident probability amplitudes, respectively.

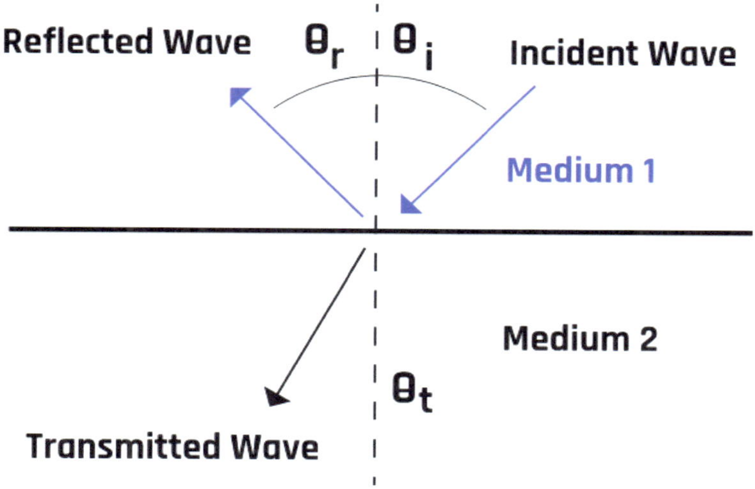

Figure 1-5. Ultrasound reflection at the boundary between two tissues with different acoustic impedances.

Reflections can also be classified into two categories: specular and diffuse, as illustrated in Figure 1-6.

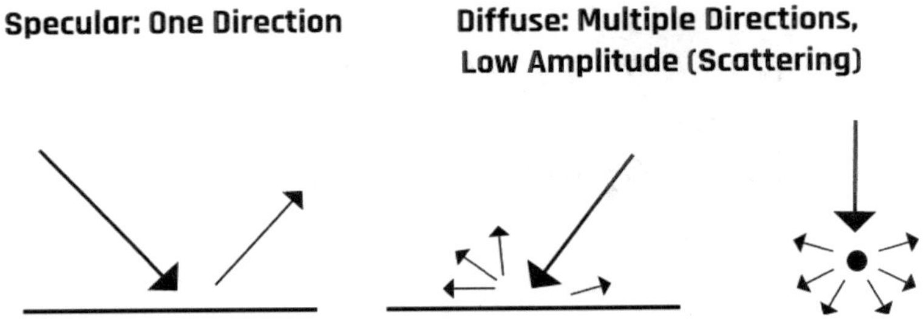

Figure 1-6: Different types of ultrasound reflections.

The ultrasound beam that succeeds in penetrating the boundary layers or interface is called the *transmitted wave*. The transmission coefficient is mathematically expressed in the following form (refer to Figures 1-5 and 1-7):[2]

$$\frac{P_t}{P_i} = \frac{2Z_2 \cos\theta_t}{Z_2 \cos\theta_i + Z_1 \cos\theta_t},$$

where P_t represents the transmission probability amplitude.

The ratio of the speed of the transmitted wave (v_2) to that of the incident wave (v_1) is related to the ratio of the sines of the angles of transmission and incidence, a relationship called Snell's law:

$$v_2 \cos\theta_t = v_1 \cos\theta_i.$$

2. Chan VWS. Ultrasound Imaging for Regional Anesthesia. 2nd ed. Toronto, ON: Toronto Printing Company; 2009. 202 p.

When the waves are reflected from a perfectly flat surface or boundary, the reflected waves tend to be uniformly parallel to each other. In contrast, they tend to be diffuse for rough surfaces. This phenomenon is commonly observed as a "mirage" when driving on a hot summer day, and the road appears to have a wet surface that disappears as one gets closer. This leads to two different kinds of reflections, specular and diffuse, which are illustrated in Figure 1-6.

The transducer picks up the reflected waves and converts the echoes into images. The strength of the echoes depends on the acoustic impedance between the two tissues through which the waves pass. Typically, boundary reflections occur on blood vessel walls and organ boundaries.

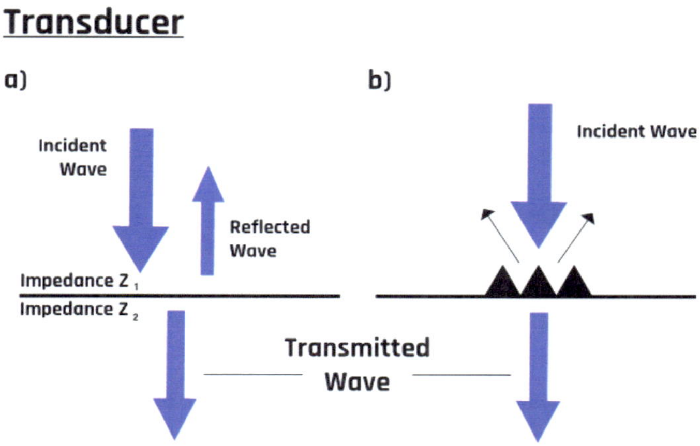

Figure 1-7: The ultrasound beam from the transducer undergoes reflection, refraction, and absorption as the beam penetrates the tissue.

1.8.2 Refraction

When an ultrasound beam strikes a tissue boundary obliquely, the transmitted component of the beam undergoes a change in direction. This change is due to the differences in the velocities of the incident and transmitted beams. This bending process, called *refraction*, is illustrated in Figure 1-8 and is often related to the formation of artifacts during ultrasound image acquisition.

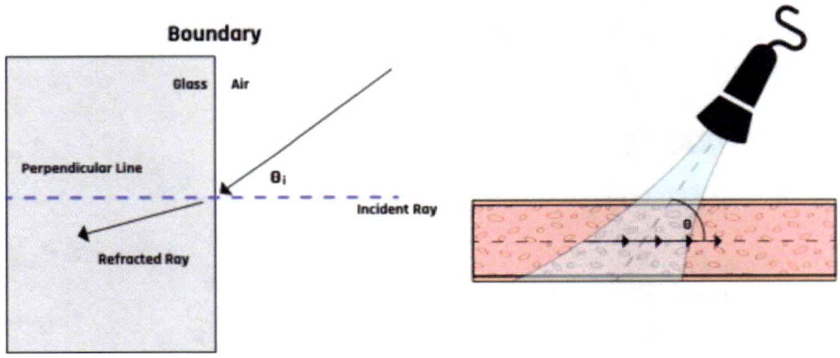

Figure 1-8: Ultrasound refraction at surface boundaries.

In ultrasound imaging, refraction can result in the formation of artifacts such as double image artifacts, as shown in Figure 1-9. This artifact is caused by the differential refraction of the ultrasound beam while passing through the relatively different echogenic tissues, such as muscle and fat tissues, and the difference in velocities in those tissues.

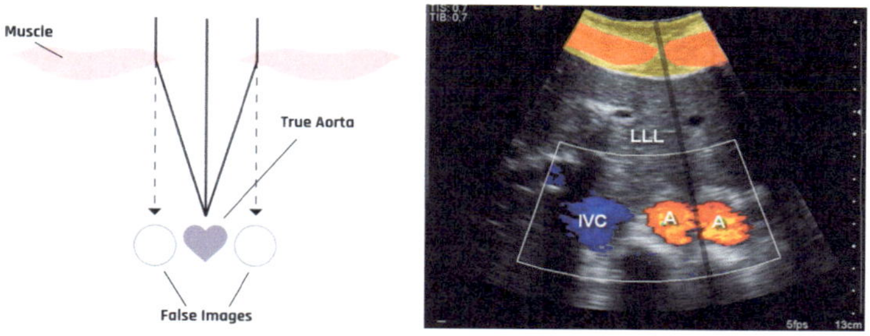

Figure 1-9: The figure on the left illustrates how a double image artifact is formed, and the one on the right is the actual double image artifact of an aorta on an ultrasound image. Aorta duplication artifact by Nevit Dilmen licensed under CC BY-SA 3.0

1.8.3 Scattering

Inside the human body, scattering is mostly a result of small changes in the density, compressibility, and absorption properties of the tissues. Ultrasound scanners detect these scattered waves to show the backscattered signal in the form of images. Scattered waves are primarily caused by red blood cells, or erythrocytes, due to their relatively small diameter of 6–8 μm, compared to the commonly used ultrasound wavelength of about 0.8 mm. For ultrasound frequencies below 20 MHz, the backscattering signal from blood is about 10 to 27 dB lower than from the surrounding tissue.[3] This difference makes it possible to image the blood flow inside the tissue.

3. Jensen JA. Estimation of blood velocities using ultrasound: A signal processing approach. New York: Cambridge University Press; 1996. 317 p.

1.8.4 Absorption

As the ultrasound beam traverses the tissue structures, it also loses energy through absorption. The lost energy varies depending on the tissue's characteristics and the ultrasound wave's frequency. For example, bones absorb more ultrasound energy than soft tissue does. Absorption of ultrasound in tissue is frequency dependent—it increases with increasing frequency.

1.8.5 Attenuation

As the ultrasound moves through tissues, some of the ultrasound energy is lost due to absorption through heat, reflection, refraction, and scattering. The beam weakens with increased depth into the tissue, increasing acoustic impedance mismatch. Another factor is the presence of air bubbles inside the tissue, which tend to form virtually impenetrable barriers to ultrasound. Attenuation becomes higher not only with increasing distance from the transducer but also because of the heterogeneity caused by acoustic impedance mismatch as well as the higher frequency of the transducer, as illustrated in Figure 1-10. This is because air has a higher resistance to ultrasound propagation than fluids.

The intensity (I_x) of an ultrasound beam at tissue depth x can be estimated using Beer's law:

$$I_x = I_0 e^{-\mu x},$$

where I_0 is the incident intensity at the tissue surface and μ is the intensity attenuation coefficient. Of this attenuation, absorption contributes about 60–80%.[4]

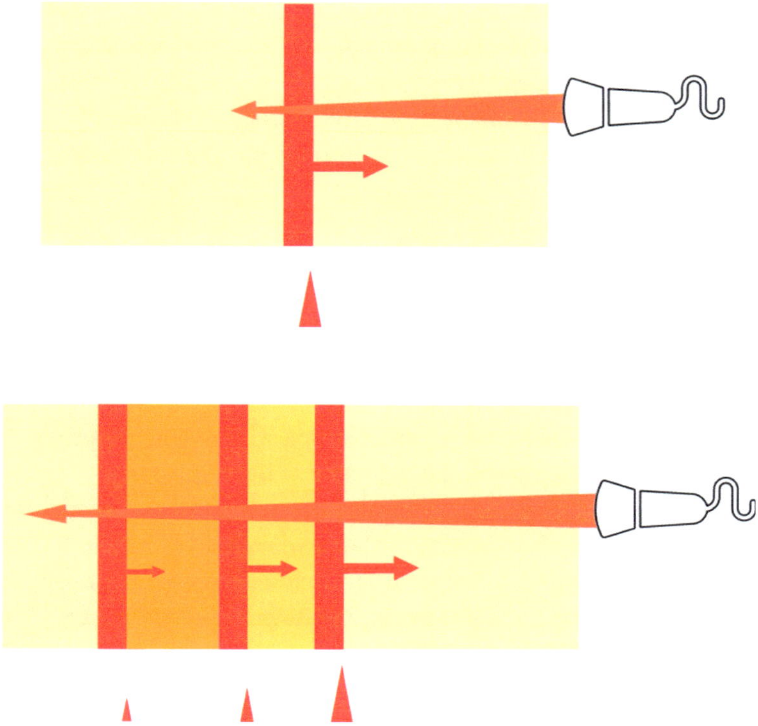

Figure 1-10: An illustration of attenuation at multiple tissue boundaries.

Attenuation increases with increasing gas and fat. The higher the tissue density (or impedance), the lower the reflection. For example, blood has an attenuation coefficient value closer to

4. Ter Haar G. Ultrasonic imaging: Safety considerations. Interface Focus. 2011 Aug 6;1(4):686–97. doi: 10.1098/rsfs.2011.0029. Epub 2011 May 25. PMID: 22866238; PMCID: PMC3262273.

0.20 dB/MHz.cm, while the typical value for bone is around 20.0 dB/MHz.cm.[2]

Attenuation generally increases linearly with increasing frequency among different body tissues. Fluid-filled structures have much lower attenuation than solid structures. Hence the transmitted pulse from a fluid-filled structure is usually more substantial than that from passing through an equivalent amount of solid tissue.

1.8.6 Diffraction

The ultrasound beam spreads out with distance from the transducer as it passes through the tissue, causing diffraction, as shown in Figure 1-11. This results in the reduction of beam intensity.

This diffraction pattern is highly dependent on the shape and size of the transducer relative to the wavelength of ultrasound. This phenomenon causes a decrease in the intensity of the ultrasound beam. To achieve a parallel beam, the diameter of the crystal face is designed to be approximately 10 to 20 times the wavelength of ultrasound.

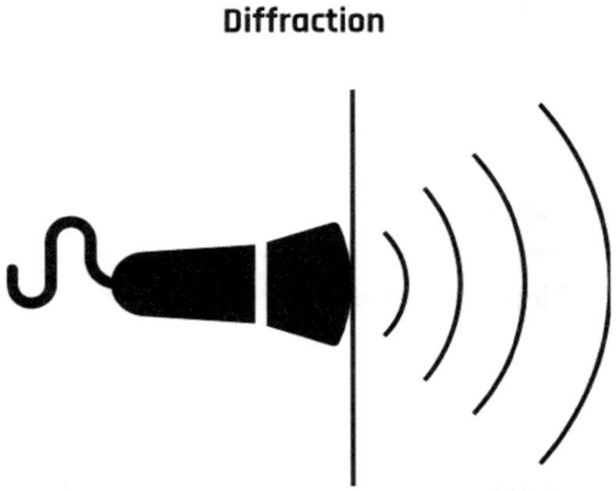

Figure 1-11: Schematic of an ultrasound diffraction.

1.8.7 Acoustic Impedance

Acoustic impedance (Z) is the resistance of a tissue to the passage of ultrasound. It depends on the density of the tissue (ρ) and the velocity of propagation (v) of the ultrasound through the tissue: $Z = \rho v$. If the density is in kg/m^3 and the velocity is in m/s, then the specific acoustic impedance is expressed in the unit of rayl (Ry), which is equivalent to 1 $kg/(m^2s)$. A typical value of acoustic impedance is 0.0004 rayls for air, 1.65 rayls for the liver and blood, and around 5 rayls for the bone.[5]

The greater the difference between two tissues (media), the more the ultrasound is reflected and the lesser the transmission through the tissue. For example, more ultrasound beams will be reflected at soft tissue / bone and soft tissue / air interfaces than at soft tissue / blood interfaces. The difference in acoustic impedance is called *acoustic impedance mismatch*. A greater acoustic mismatch results in a greater reflection and a lower transmission.[6]

1.9 Self-Assessment

1. What are the three different classifications of

5. Goss SA, Johnston RL, Dunn F. Comprehensive compilation of empirical ultrasonic properties of mammalian tissues. J Acoust Soc Am. 1978 Aug;64(2):423–57. doi: 10.1121/1.382016. PMID: 361793.
6. Goss SA, Johnston RL, Dunn F. Comprehensive compilation of empirical ultrasonic properties of mammalian tissues. J Acoust Soc Am. 1978 Aug;64(2):423–57. doi: 10.1121/1.382016. PMID: 361793.

sound waves?
2. To which frequencies do each classification of sound waves correspond?
3. What different processes can ultrasound waves undergo as they pass through the body and tissues?
4. How is acoustic impedance defined?

1.10 *Further Readings*

1. Nelson TR, Fowlkes JB, Abramowicz JS, Church CC. Ultrasound biosafety considerations for the practicing sonographer and sonologist. J Ultrasound Med. 2009 Feb;28(2):139–50. doi: 10.7863/jum.2009.28.2.139. PMID: 19168764.
2. Jensen JA. Estimation of blood velocities using ultrasound: A signal processing approach. New York: Cambridge University Press; 1996. 317 p.
3. Ter Haar G. Ultrasonic imaging: Safety considerations. Interface Focus. 2011 Aug 6;1(4):686–97. doi: 10.1098/rsfs.2011.0029. Epub 2011 May 25. PMID: 22866238; PMCID: PMC3262273.
4. Goss SA, Johnston RL, Dunn F. Comprehensive compilation of empirical ultrasonic properties of mammalian tissues. J Acoust Soc Am. 1978 Aug;64(2):423–57. doi: 10.1121/1.382016. PMID: 361793.

2. Ultrasound Instrumentation

2.1 Learning Objectives

After reviewing this chapter, you should be able to do the following:

1. Describe the ultrasound instrument components and their functions.
2. Identify the different types of ultrasound probes and their uses in clinical practice.
3. Understand the different types of ultrasound transducers and their characteristics.
4. Understand the importance of probe selection and placement for optimal image acquisition.
5. Describe the challenges of interpreting ultrasound images and how medical professionals address them.

2.2 Introduction

Before learning how to use ultrasound units to assess various anatomical and physiological features, it is essential to learn about the physics behind ultrasound machines. Ultrasound units send ultrasonic waves from a transducer that get reflected from the tissues and are then received by the transducer. The computer processes this information, and images are generated. Transducers utilize the piezoelectric effect to convert the electric pulses to sound waves and vice versa for this image processing. As with any other medical imaging device, ultrasound images are not artifact-free. Multiple different artifacts are associated with ultrasound imaging.

2.3 Ultrasound Machine Components

As shown in Figure 2-1, an ultrasound machine includes the following major components:

- Display—a screen that shows images from the ultrasound scans.
- Keyboard—a key panel for data input and measurement display.
- Central processing unit—a unit that processes signals from and to the transducer.

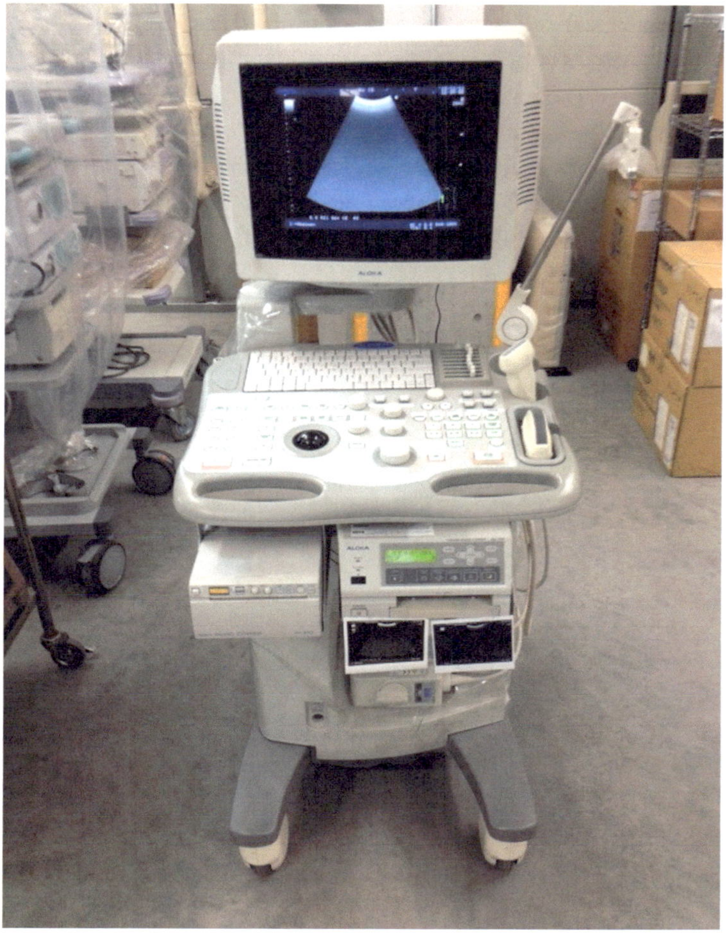

Figure 2-1: A typical ultrasound machine. ALOKA SSD-3500SV by Kitmondo Marketplace licensed under CC BY 2.0

- Pulse controls—dials and controls that are used to change the amplitude, frequency, and duration of ultrasound pulses.
- Transducer—a probe that generates ultrasound waves and detects reflected echoes. It contains piezoelectric materials, which vibrate due to echo pulses from the tissue. The

transducer relies on the piezoelectric effect. This piezoelectricity is amplified and transmitted to the display, where it is converted into an image form.

- Amplifier—a unit that increases the size of the electrical pulses coming from the transducer after an echo is received. The amount of amplification is controlled by the gain control knob, which allows the user to adjust the gain to the required depth within the body.
- Storage device (not shown)—a digital device that stores images for later use.
- Printer—a unit that prints images from the displayed data.

Figure 2-2 shows a generalized schematic of ultrasound machines' mode of operation.

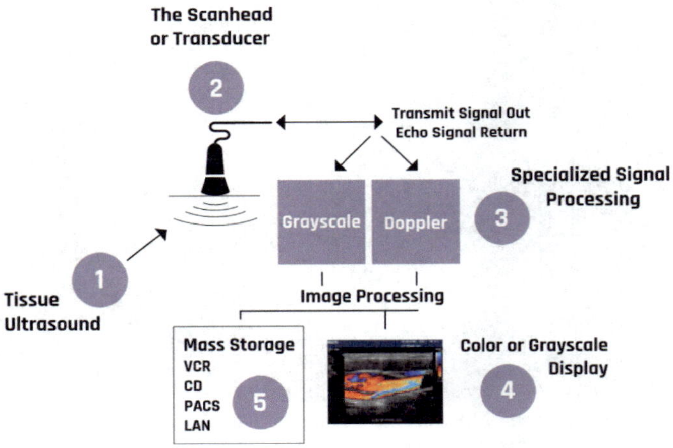

Figure 2-2: Schematic of the mode of operation of ultrasound machines.

More details about the operations of the components of the ultrasound machine are discussed in the sections below.

2.4 Piezoelectric Effect

In the 19th century, Pierre and Jacques Curie discovered that some materials generate electric potentials in response to mechanical deformation (material shrinks or expands) or stress (the substance is squeezed or stretched)—a phenomenon called the *piezoelectric effect*, which is illustrated in Figure 2-3. Conversely, the same materials change their shapes when an electric field is applied.

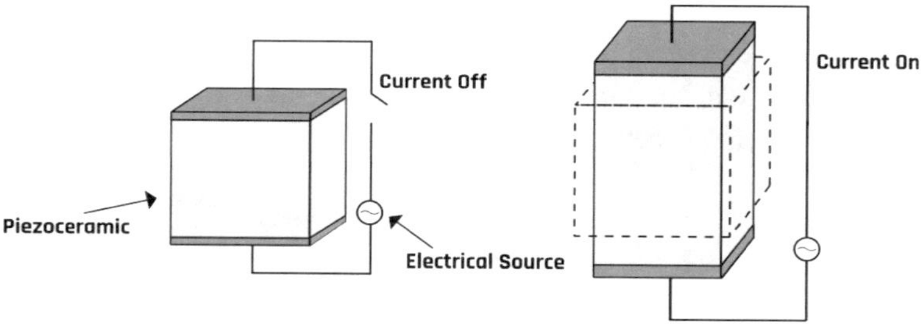

Figure 2-3: An illustrative diagram of the behavior of piezoelectric materials used in ultrasound probes.

Examples of such materials include silicon oxide, potassium sodium tartrate, barium titanate, and lithium niobate. Bones, tendons, skin, and some man-made polymeric materials can also exhibit the piezoelectric effect. These materials bend in different ways depending on the frequency and their shape, which can result in different vibration modes. The modes are the basis for developing transducers that operate at different frequencies.

2.5 Transducer Characteristics

A typical ultrasound probe with its components is shown in Figure

2-4. The vibration of crystals in an ultrasound machine's transducer generates ultrasound, and the transducer can detect the echoes and convert them to electrical signals. A transducer is composed of piezoelectric crystals, which respond to pressure to generate an electric current. The alternating current causes the piezoelectric crystals to vibrate at a desired frequency corresponding to ultrasound waves. The produced ultrasound beam is directed into the tissues by moving the transducer and changing the angle of incidence of the ultrasonic beam. Conversely, when an electric current is applied to the crystals, its shape changes with polarity, producing electrical signals from echoes that are processed to generate a display. Hence the crystals act as transmitters (for a short time) and receivers (most of the time).

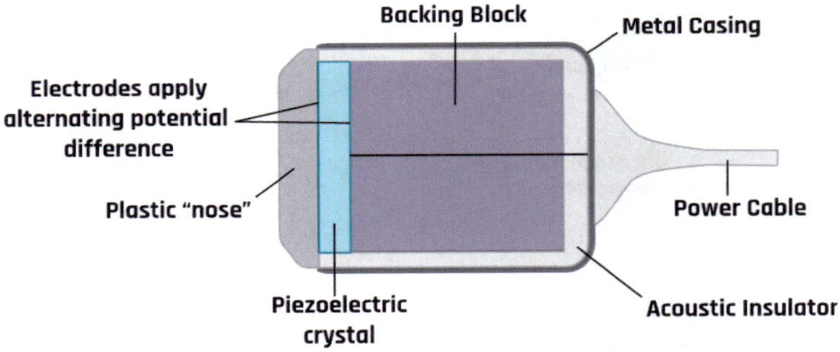

Figure 2-4: Components of an ultrasound probe.

Since the air between the tissue and the transducer inhibits the propagation of the ultrasound beam, a conducting gel is usually applied between them.

2.6 Image Display and Grayscale

Figure 2-5 shows a block diagram of an ultrasound imaging system. Two modes are essential in the formation of an ultrasound image. These are the *transmission modes* that convert an alternating current into mechanical pressure waves. The backscattered pressure waves are picked up by the *receiving mode*, which converts them into electrical signals. The ultrasound waves that get fully transmitted through any tissues or structures do not produce echoes and appear dark. For example, all fluids appear echo-free and black on the ultrasound image.

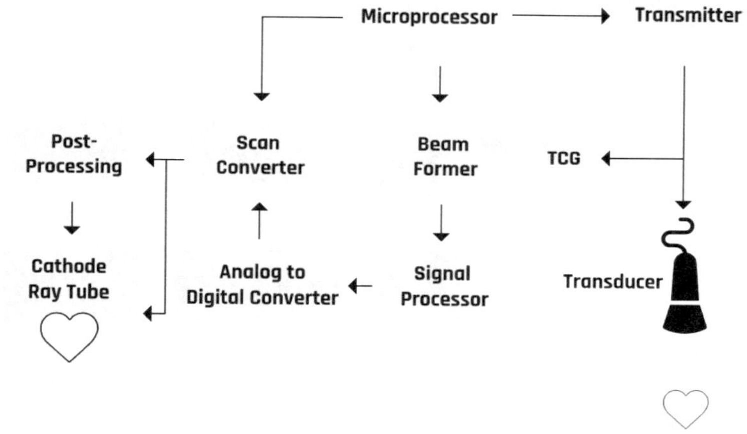

Figure 2-5: Block diagram of an ultrasound imaging system.

2.7 Pulse-Echo Imaging

The transducer generates pulses and detects backscattered energy from the tissue boundaries, as shown in Figure 2-6. The length of delay between the transmitted and received pulses is used to determine the depth of the tissue boundary or organ under examination.

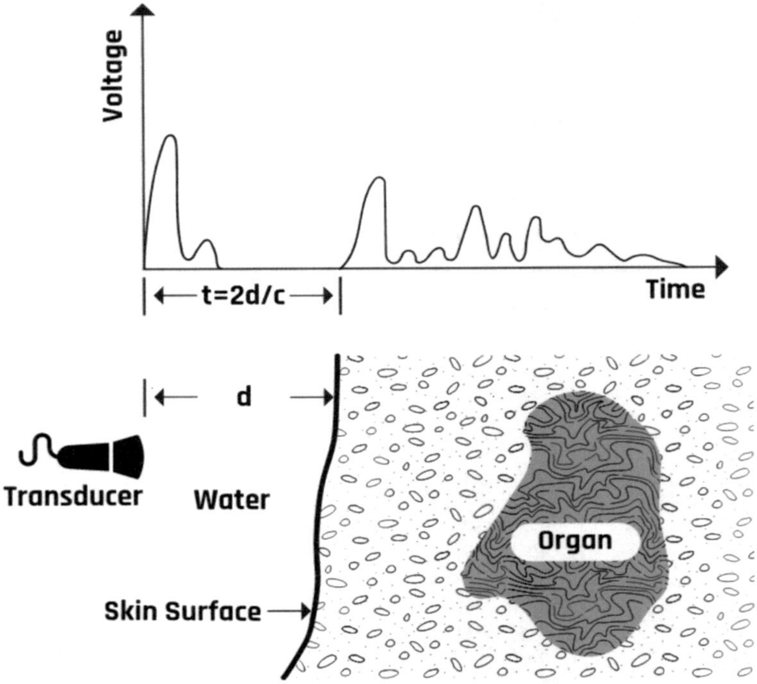

Figure 2-6: An illustration of the pulse-echo imaging operation.

These piezoelectric signals from the crystals are amplified and converted into a gray or white color on the ultrasound image via a computer program. The difference in tissue reflectivity allows us to see individual structures. When ultrasound hits a dense object, it is wholly reflected, forming a posterior acoustic shadow (a bright and echogenic image). This is because no ultrasound is transmitted, creating an echo void. The computer can calculate the tissue's depth by measuring the time between when the wave was sent and when an echo was detected.

2.8 Image Acquisition and Analysis

There are several assumptions about sound waves in ultrasound system operation. These assumptions, when violated, result in the formation of image artifacts, which often make it difficult to distinguish between real and fictitious features in an image. These assumptions are

- the ultrasound beam travels in a straight line with a constant rate of attenuation;
- the ultrasound waves travel directly to a reflecting tissue and back;
- the speed of ultrasound in all soft tissue is exactly 1540 m/s;
- reflections arise only from structures positioned in the beam's main axis;
- the ultrasound beam is infinitely thin, with all echoes originating from its central axis;
- the strength of a reflection is accurately determined by the characteristics of the tissue creating the reflection; and
- the depth of a reflector is the time taken for sound to travel from the transducer to the reflector and return.

2.9 Artifacts and Errors Associated With Image Acquisition

Artifacts are fictitious portions of images (distortions of the actual anatomy of the tissue). Improper scanning techniques may cause some artifacts, while others result from the physical limitations of the instrument. Such artifacts can be explained using the properties of the ultrasound waves, their propagation through tissue, and the assumptions used in image processing. Typical artifacts include shadowing, beam width, side lobe, reverberation, comet tail, ring down, mirror image, and refraction. They are formed primarily due to multiple echo paths or velocity errors and attenuation errors.

2.9.1 Shadowing Artifact

Highly reflective or attenuating tissues reduce the ultrasound beam intensity inside the tissues, leading to obscured images close to or behind them. This shadowing artifact results from refraction that causes echoes to appear darker, like a shadow, due to the ultrasound beam's decreased amplitude, intensity, and power.

Shadows occur when an ultrasound beam cannot pass through an area deeply due to the presence of a strongly reflecting or attenuating tissue. Figure 2-7 shows acoustic shadowing caused by the stones in the gallbladder. The shadows occur in regions of high acoustic impedance mismatch, such as soft tissue / gas or soft tissue / bone interfaces. These shadowing artifacts prevent visualization of the accurate anatomy on a scan by covering it with an anechoic shadow. This may cause misdiagnosis of the tissue anatomy.

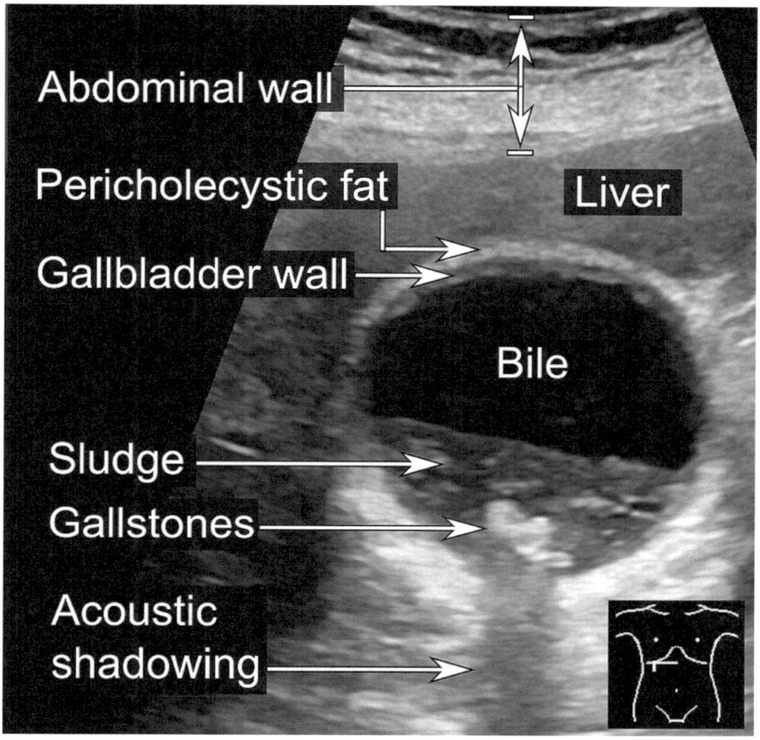

Figure 2-7: The stones in gallbladders are very bright, and they cast an acoustic shadow posterior to the stones. Ultrasonography of sludge and gallstones, annotated by Kitmondo Marketplace licensed under CC0 2.0

Other causes of shadowing artifacts are improper scanning techniques, improper settings, or poor ultrasound systems.

Some possible ways to reduce these artifacts include taking images from several angles, changing the lateral resolution, or decreasing the frequency to avoid missing information.

2.9.2 Mirror Image Artifact (a.k.a. Ghost Artifact)

Multiple reflections (reverberation) often occur in regions with high impedance mismatches, such as air/fluid or flesh/bone interfaces. The multiple reflections duplicate a true reflector when the waves from a highly reflective surface are redirected toward a second structure. The redirected waves form a replica of the original structure, which appears on the image as a second structure.

Mirror image artifacts occur in both grayscale and color Doppler imaging. The true reflector and the artifact are equidistant from the mirror plane located between the two, as shown in Figure 2-8. This violates the assumptions that (1) ultrasound waves travel in a straight line and (2) waves travel directly to a reflecting tissue and are reflected directly back to the transducer.

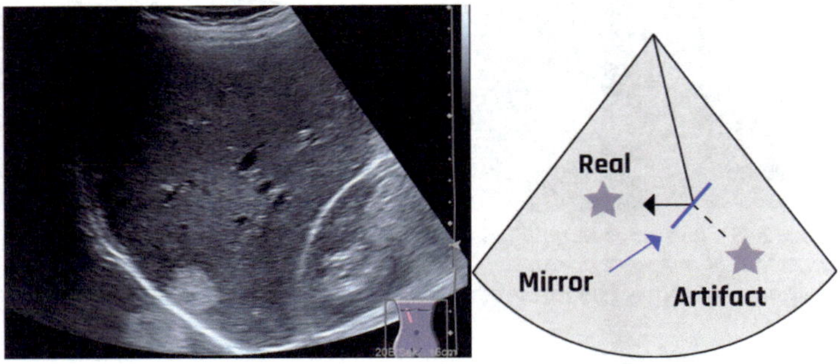

Figure 2-8: A mirror image artifact appears as a symmetric image with less intensity than the actual image on the opposite side of the baseline. The mirror image artifact is apparently outside the liver. Leberhaemangiom mit Spiegelartefakt 56M – US – 001 by Hellerhoff licensed under CC BY-SA 4.0

A ghost artifact can develop on a color Doppler image when multiple reflections occur beyond borders. Ghost or mirror image artifacts can be reduced by decreasing the overall gain or changing the beam angle. During diagnosis, mirror images should be clearly separated from actual anatomy, especially in guided needle biopsy, where samples must be taken from a specific location.

Two possible ways by which mirror image artifacts can be formed are illustrated in Figure 2-9a.

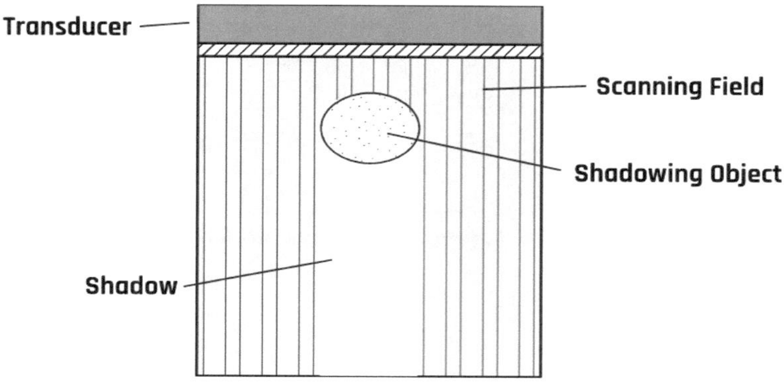

Figure 2-9a: One of the assumptions of ultrasound imaging is that the beam travels in a straight line.

Circular structures such as cysts can cause refraction, which produces shadows on the object's edge due to a mismatch in acoustic impedance at the boundary or interface. The change in direction (bending) results from the change in the propagation velocity of the ultrasound waves. A schematic of this process is illustrated in Figure 2-9b.

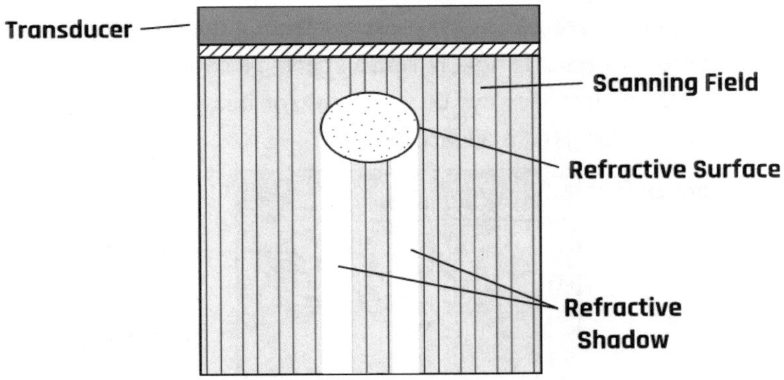

Figure 2-9b: Schematic of the formation of a ghost artifact.

2.9.3 Side Lobe Artifact

Beams generated from the edges of a single-element transducer tend to spread from the primary beam, as shown in Figure 2-10. These lobe beams can be reflected into the primary beam, adding energy to the beam's main axis. These artifacts violate the assumption that all reflections occur in the path of the beam's main axis.

This duplication of the true anatomy with false reflection results from strong reflections that return to the transducer. Since the machine assumes all echoes to be coming from a true anatomic structure, incorrect images are displayed together with the correct ones.

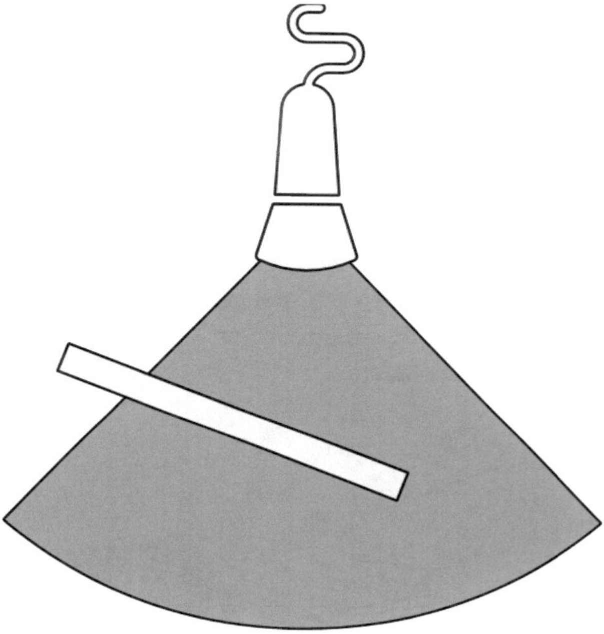

Figure 2-10: Image illustration of a side lobe effect.

A rapidly oscillating ultrasound beam produces multiple side lobe echoes that appear on the display as a curved line equidistant from the transducer. The two most important features that distinguish a side lobe artifact from an anatomical structure are that it is equidistant to the transducer along its length and it passes through anatomical structures.

The side lobe artifact is corrected by imaging the structure in multiple directions. The artifact will not appear in all viewing directions.

2.9.4 Grating Lobe Artifact

While the assumption is that all reflections are in the path of the beam's main axis, as the beam is projected from the transducer into the tissue, some of the ultrasound waves spread outward, as shown in Figure 2-11. The presence of a strong reflector along the path of the diffracted waves produces echoes that are misinterpreted as being along the beam's main axis. The grating lobe artifact generally appears weaker than the true reflector. This obscures the actual anatomy with a false reflection.

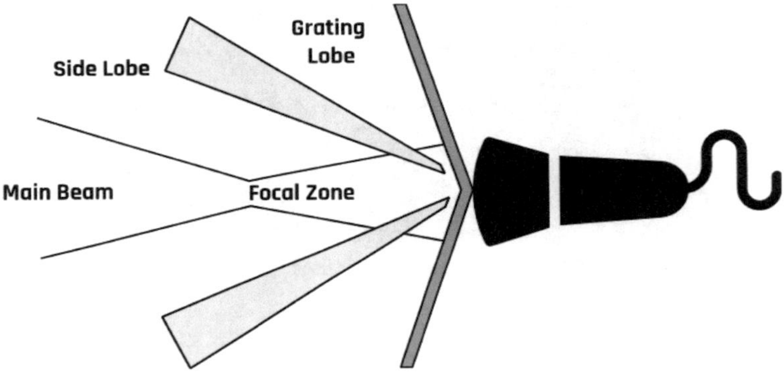

Figure 2-11: An ultrasound energy beam that deviates from the central beam results in side or grating lobes, which, upon encountering a reflector, can produce echoes that return to the transducer. The machine erroneously assumes the echoes are due to reflection from the main beam axis and displays the image in the wrong location.

The artifact is corrected by taking multiple views of the structure under examination. An artifact will not appear in all views.

2.9.5 Multipath Artifact

Multipath artifacts occur when the primary ultrasound beam

reflects off anatomy at an angle such that a part of the echo returns to the transducer and at the same time, another echo also reaches the transducer after reflecting off a second boundary.

The echo from the secondary reflector takes a longer path and hence longer time to get back to the transducer. Since ultrasound machines measure the depth based on the time between the transmitted signal and the received echo, the machine will perceive a longer time and depth and position the image on the wrong spot, as shown in Figure 2-12.

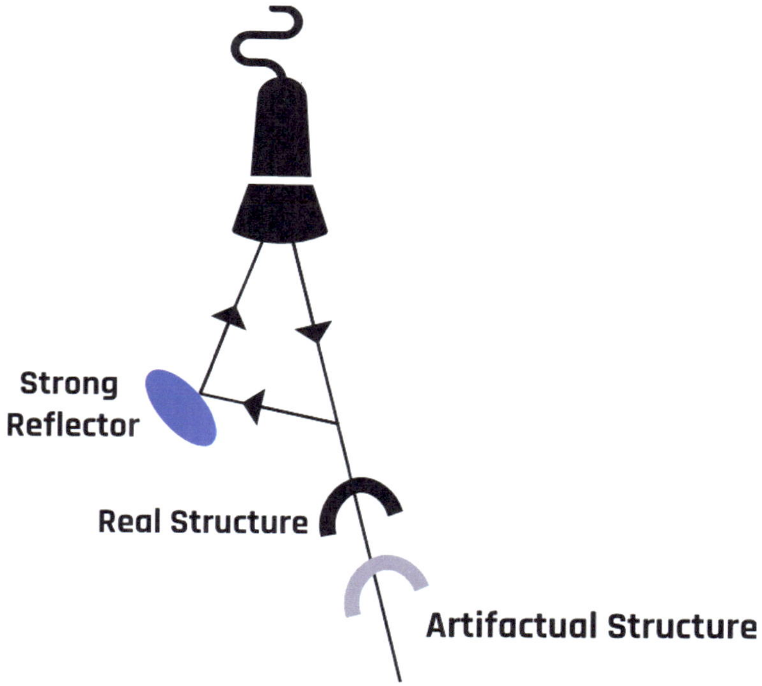

Figure 2-12: A schematic of multiple reflections from two surfaces.

In this case, the assumption that the ultrasound beam travels directly to the reflector and back to the transducer is violated. This phenomenon creates what is called a propagation path error. These artifacts give rise to the incorrect axial location of an object due to longer path lengths.

Multipath reflections may form images that appear deeper or misplaced. The problem can be reduced by taking multiple views at different angles.

2.9.6 Curved/Oblique Reflectors

An ultrasound beam incident on a curved or oblique boundary is reflected in various directions. Some of the reflected waves are directed away from the transducer. This reflection is similar to the scattering process. In this case, reflectors do not appear on the image due to longer path lengths and increasing attenuation.

Figure 2-13 illustrates the possible reflections from an oblique surface in various directions. This observation contradicts the assumption that an ultrasound pulse travels directly to a reflecting boundary surface and back to the transducer.

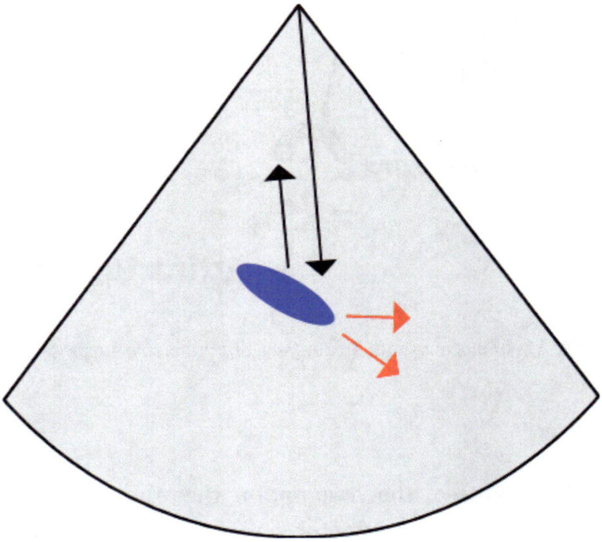

Figure 2-13: Reflections from an oblique surface.

The strength of the echo received by the transducer is less than the expected intensity, which gives false brightness, missing reflections, or an improper location of the anatomic structure. These artifacts are associated with weak, too-bright echoes or improperly located structures. This artifact can be reduced by changing the transducer angle or by using a large footprint.

2.9.7 Beam Width Artifact

While it may be convenient to assume that the beam width stays approximately equal to the transducer size, the ultrasound beam actually spreads out as it moves away from the transducer, as Figure 2-14 shows. Due to this divergence, the echoes generated from the edge of the beam appear to be coming from the center of the beam. The artifacts are most apparent when most of the beam travels through the fluid and part of it interacts with adjacent soft tissue, as shown in Figure 2-14.

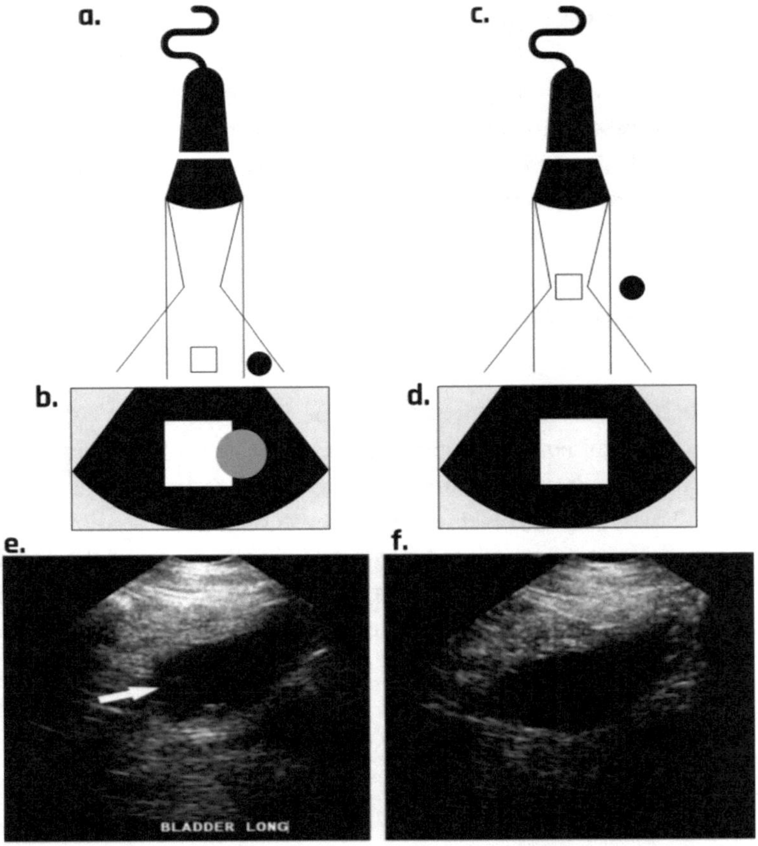

Figure 2-14: Beam width artifact can cause an image to appear in the wrong location due to side echoes that are erroneously interpreted by the machine as part of the central beam.

2.9.8 Reverberation Artifact

While the ultrasound machine is based on the assumptions that (1) sound travels in a straight line, (2) all echoes are parallel to the transducer axis, and (3) sound waves travel at 1540 m/s in soft tissue, the ultrasound echoes may be reflected repeatedly between two highly reflective surfaces that are parallel to the primary ultrasound beam. This causes reflections that oscillate between the tissue and the transducer. The artifacts appear on the image as multiple stairways that are equally spaced apart from one another and tend to increase with increasing tissue depth.

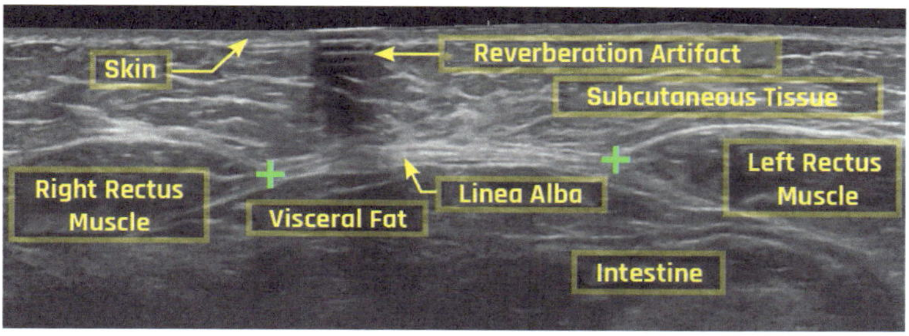

Figure 2-15: Reverberation artifacts are characterized as multiple horizontal lines that are equidistant from one another, and they are more pronounced as the intensity decreases with the depth. Ultrasonography of diastasis recti – annotated by Mikael Häggström, M.D. licensed under CC0 1.0

Figure 2-15 shows an image with multiple reflections, indicated by arrows at the top. The first bright line at the top close to the transducer is the only real line image. The other bright images below the actual reflector are artifacts. Another way reverberation artifacts can occur is when the transducer behaves as another reflecting surface such that the returning echoes are rereflected back into the tissue-reflecting structure, resulting in the formation of an identical artifact located at twice the distance from the transducer.

Because of attenuation, each image formed due to subsequent echoes is weaker than the first, as shown in Figure 2-15. The artifact can be prevented by moving the transducer probe at various angles to see an area covered by the artifact.

2.9.9 Comet Tail Artifact

There are various possible causes of comet tail artifacts. These are

- multiple reflections between very closely spaced reflectors;
- small calcifications and the presence of metal objects;
- vibrating small, highly reflective surfaces such as air bubbles; or
- the presence of reflectors in a medium with high velocity.

The artifact typically appears as one or multiple solid bright "tails" parallel to the axis of the primary sound beam, as shown in Figure 2-16. The pattern may differ depending on the size, shape, and composition of the reflecting tissue structure and the scan orientation and distance from the transducer.

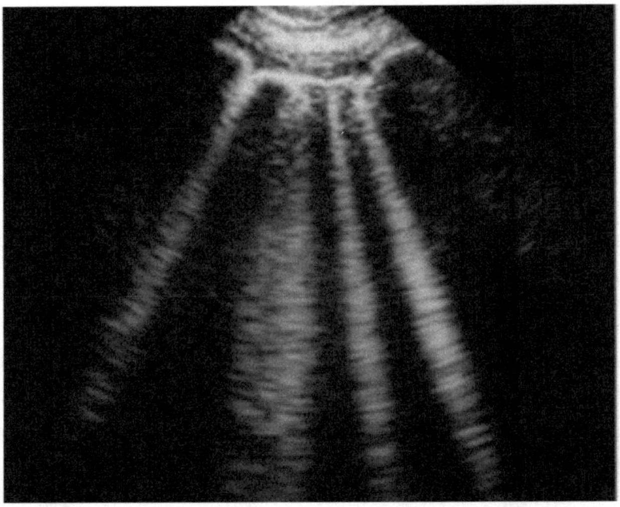

Figure 2-16: Comet tail artifact seen in the intercostal space. Interstitial syndrome adn the lung rockers by Daniel A Lichtenstein licensed under CC BY 2.0

This artifact helps diagnose or rule out pneumothorax. If the pneumothorax is present, the air within the pleural space hinders the propagation of ultrasound waves, thereby preventing the formation of comet tail artifacts.

The major problem is that the tails cause significant attenuation such that the beam becomes significantly weak and cannot reach deeper regions of the tissue. These tails prevent the scan from imaging the underside of the reflecting structure. The artifact can be prevented by performing multiple scans at different angles to view the area obscured by the tail.

2.9.10 Propagation Speed Error Artifact

One of the other assumptions of ultrasound machine operation is that the speed of sound in soft tissue is exactly 1540 m/s. However, sometimes the waves may propagate through a medium at a speed other than that of a soft tissue. This produces the correct number of reflectors, which appear at incorrect depths. For speeds greater than 1540 m/s, the depths are underestimated due to the short transmission-reception time of the beam from the transducer to the reflector and back to the transducer. When the sound travels at a speed slower than 1540 m/s, the distances of the reflectors are overestimated.

These artifacts can be reduced by changing the beam's angle, which may help minimize the difference in propagating speed. Speed error artifacts cannot be prevented entirely. It should be remembered that the artifact causes incorrect placement of the reflectors on the image display.

Care must be taken to identify incorrect placements to avoid misinterpreting the image. This is often achieved by taking images from different angles. If the placement of the reflectors cannot be duplicated at different angles, then the image is an artifact. These speed errors can make the image appear "split" or "cut." The speed error degrades the quality of an image that relies on resolution, such as the differentiation of lesions and cysts and guided biopsy.

2.9.11 Resolution Artifact

Two assumptions of the ultrasound machine are that echoes come from the main axis of the beam and that sound travels in a straight line. However, very small structures between the beamlines degrade the image detail and produce misleading echoes. The artifact can be prevented by having a higher spatial resolution or line density. Low line density per frame produces poor detail in images. Another cause is excess gain, which tends to create or obscure information and reduce lateral resolution.

The resolution can be corrected by choosing the proper gain or focal zone setting to reduce erroneous echo contrast. An appropriate setting reduces the grainy pattern.

2.9.12 Near Field Clutter

Near field clutter is an artifact that arises from multiple noise sources. Any acoustic noise near the transducer may cause high-amplitude oscillations of the piezoelectric crystals in the transducer. It involves the near field and may hinder the identification of structures that are close to the transducer. These oscillations cause the artifacts to appear and disappear. In general, artifacts change their appearance and appear or disappear depending on the view.

The artifact occurs due to acoustic noise near the transducer, resulting from high-amplitude oscillations of the piezoelectric elements. The appearance of additional echoes from high-amplitude reflections from the transducer in the near field can overshadow the weaker echoes from true anatomic structures. Nevertheless, when viewed from multiple angles, the real tissue structures will remain constant, while the artifacts will appear and disappear as the scan angle changes.

2.9.13 Ring Down Artifact

This artifact is caused by small gas bubbles, which produce reflections after the transducer receives the initial reflection. The transducer sees the echoes as though they are coming from structures in the deeper part of the tissue. The pocket of fluid and air continuously resonates, reflecting ultrasound and creating a hyperechoic region. This process commonly occurs in regions with air bubbles and water.

The artifact generally appears as very bright and continuous parallel bands extending to the image's bottom. Its brightness makes it hard to view the region beneath it. The effect can be reduced by moving the scanning beam at different angles.

2.9.14 Enhancement Artifact

This artifact is the opposite of the shadowing effect due to low attenuation. It is common in fluid-filled structures such as the gallbladder, the urinary bladder, or cysts due to excessive brightness. The primary causes are improper ultrasound settings and scanning techniques.

2.9.15 Focal Enhancement and Banding Artifact

Focal enhancement is formed due to attenuation effects, improper ultrasound settings, and improper brightness, which cause the sound waves to weaken as they propagate in the medium. This is because the amplitude decreases.

2.9.16 Refraction Artifact

The change in the propagation speed of ultrasound due to differences in impedance causes the transmitted beam to change its direction after passing through the boundary. When the beam moves from one medium to another with a higher (or lower) impedance, the transmitted beam is refracted toward (or away from) the normal at the point of incidence.

This phenomenon occurs when a beam hits the interface obliquely, creating separate beams with different propagating speeds, contrary to the assumption that sound travels in a straight path. This can create a duplication of the anatomic features. For example, refraction may cause a single feature to appear as a double feature.

2.9.17 Slice Thickness Artifact

A propagating artifact appears when the beam dimension is far greater than the reflector size. Another possible cause is improper elevation resolution. The artifact "creates" debris that appears in cyst images, which can lead to false diagnoses. This is because the image plane is neither extremely thin nor uniform, as assumed. Using tissue harmonic imaging and thinner transducer arrays would reduce the problem.

2.10 Transducer Arrays

Transducers are available in various shapes and sizes depending on users' needs. As discussed earlier, a transducer consists of piezoelectric crystals—not one crystal but an array of multiple crystal elements. The various arrays and transducers are discussed below, and their schematics are shown in Figure 2-17.

2.10.1 Linear Sequential Arrays

Modern linear ultrasound transducers contain 256 to 512 elements. Their advantage is high sensitivity when directed perpendicular to the surface under examination. However, the beam cannot be steered, which limits the field of view.

2.10.2 Curvilinear or Convex Arrays

The arrays are similar to those in the linear array but are curved, which gives them the advantage of scanning a wider field of view than linear arrays.

2.10.3 Linear Phased Arrays

Phased arrays have much smaller elements than those in a linear array. Typically, they contain 128 elements to transmit and receive each data line. Linear phased arrays are typically used for viewing restricted acoustic windows.

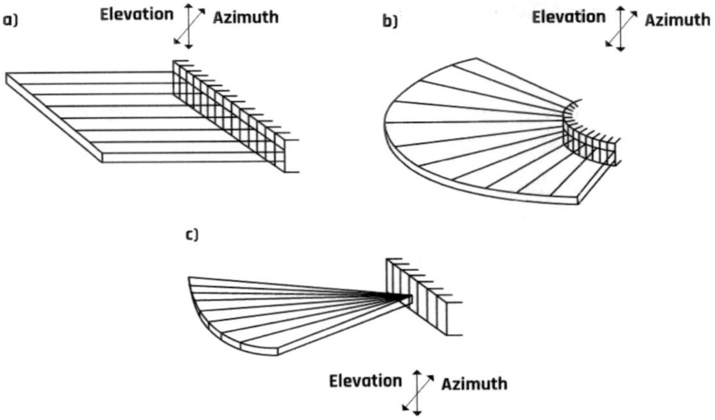

Figure 2-17: Schematics of the different types of transducer arrays: (a) linear sequential, (b) curvilinear, and (c) linear phased.

2.11 Image Display Modes

Ultrasound images are generated when the transducer transforms the reflected wave or echo from the mechanical energy of vibration into an electrical signal that is converted into an image on the display. The image can be displayed in any of the following three modes: (1) amplitude (A) mode, (2) brightness (B) mode, or (3) motion (M) mode. These are discussed in detail in the sections below.

2.11.1 A-Mode Display

The A-mode is the first form of image display. The depth (reflector-time relationship) is represented on the horizontal axis, and amplitude is displayed on the vertical axis. The A-mode measures the reflectivity at different depths under the transducer. It is commonly used for examining the eyes, the liver, and the brain. Low frequencies of 2–5 MHz are used for abdominal, cardiac, and brain scanning. Organs such as the eyes and peripheral blood vessels use a 5–15 MHz frequency.

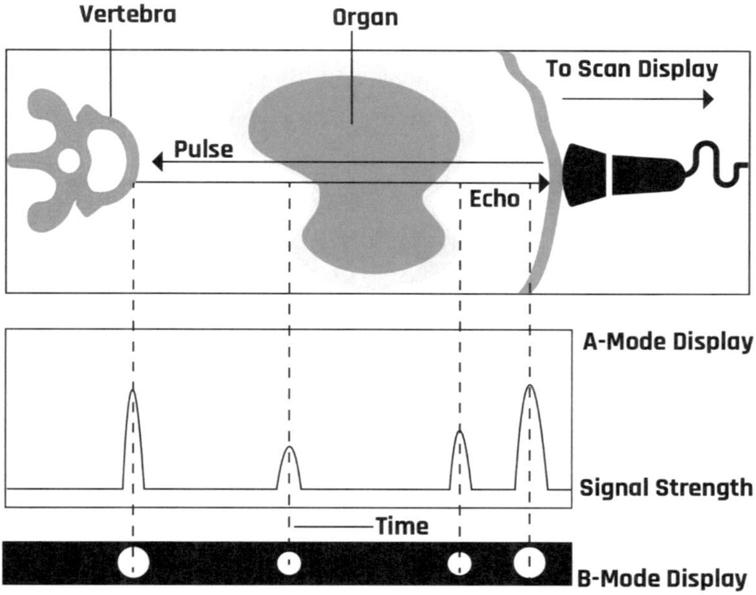

Figure 2-18: The vertical position is expressed as a voltage signal amplitude, indicating the relative amplitude of the echoes.

2.11.2 B-Mode Display

In the B-mode, the echoes are represented by bright dots. A shade of gray is assigned to each echo. The brightness of the dots indicates the strength of the echoes. The position of a dot on the screen represents the reflector distance and is determined by the transducer-reflector time relationship. Many diagnoses are made in the B-mode (in black and white) with a relatively simple probe and protocol. It finds use in the study of both stationary and moving structures. The B-mode is an electronic conversion of the A-mode and A-line information into brightness-modulated dots on the display screen, as illustrated in Figure 2-18.

The B-mode display can be used for the M-mode and 2D grayscale imaging. Modern B-mode ultrasound uses both the fundamental and the second harmonic frequencies. Harmonic imaging is most useful in patients with thick and complicated body wall structures. Figure 2-19 shows anatomical structures in the B-mode.

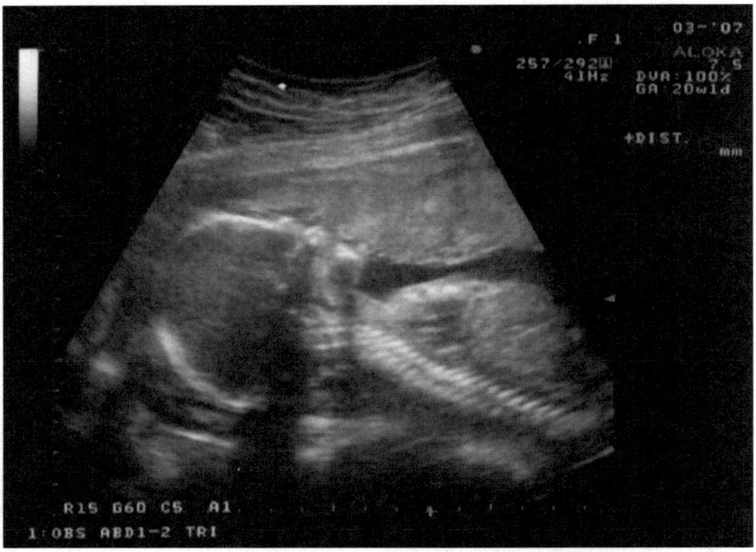

Figure 2-19: The B-mode displays anatomic structures by utilizing different gray brightness in a 2-dimensional space. Scan20semanas1 by Guimi licensed under CC BY-SA 2.5

The B-mode is also used for early intima-media thickness analysis of the carotid arteries (located in the neck) to determine the potential for lethal cardiac events. Abnormal thickening of the arterial walls of the carotid arteries is an early indicator of vascular disease throughout the body. The thicker the arterial wall, the greater the risk of heart attack or stroke.

2.11.3 M-Mode Display

In the M-mode, the motion of an object points along the transducer axis and is revealed by a bright trace moving up and down across the image. This display is commonly used to evaluate the morphology, movement, and velocity of cardiac valves and walls.

Imaging the pattern of moving cardiac structures over time constitutes M-mode echocardiography, as Figure 2-20 shows.

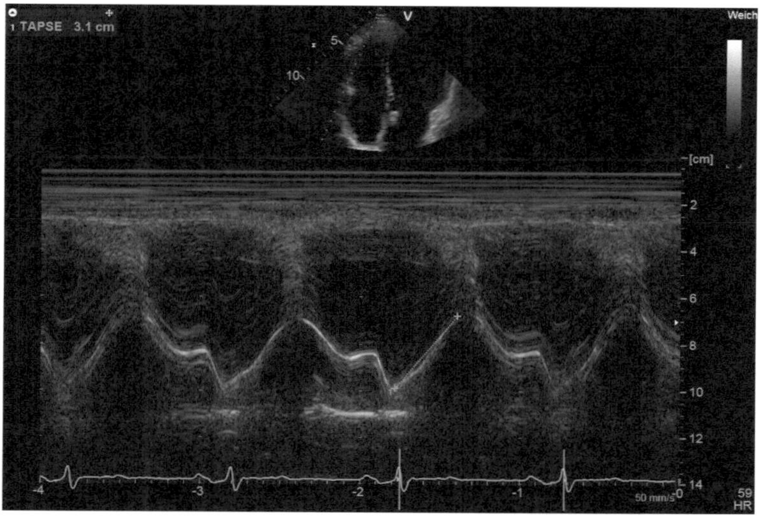

Figure 2-20: Typical example of the M-mode showing a four-chamber view of the heart. Echokardiographie M-Mode 4KB TAPSE by Wolff-BI licensed under CC BY-SA 3.0

2.12 Doppler, Color Flow, Color Power, and Duplex Ultrasound

Doppler imaging is based on the Doppler effect, which shows the relationship between velocity and frequency shift. This imaging system is mainly used to measure blood flow velocity to determine any narrowing of the arteries and assess the risk of stroke occurrence.

In the sonographic application of the Doppler effect, a typical moving source would be flowing blood, and a typical receiver would be a stationary transducer. When the source is moving away, the detected frequency is lower. Conversely, a source moving closer to the receiver would have a higher detected frequency.

For a stationary source (transducer) and a receiver (target) moving with velocity (v) at an angle (θ) relative to the direction of the incident wave of frequency (f_S) from the transducer, the *Doppler frequency* is given by

$$f_D = \frac{v \cos \theta}{c} f_S,$$

where c is the speed of sound in the tissue, which is 1540 m/s.

The transducer transmits and receives sound waves in the form of sinusoidal signal waves. The magnitude of the Doppler shift is related to the velocity of the blood cells or moving tissue, and the polarity of the shift reflects the direction of blood flow. The blood flowing toward the transducer is positive, and the blood flowing away from the transducer is negative.

The Doppler shift (Δf) is directly proportional to the velocity (v) of the blood cells, the transducer frequency (f_S), and the cosine of the angle of incidence (θ) and is inversely proportional to the velocity of sound in tissue (c = 1540 m/s). In cardiac applications, the angle of incidence in the Doppler equation is assumed to be 0 or 180 degrees.

Three modalities are currently used in Doppler echocardiography, and these are pulsed wave (PW) Doppler, continuous wave (CW) Doppler, and color flow (CF) Doppler imaging. Color flow imaging evaluates the Doppler flow information for its direction toward or away from the transducer based on the color display. A commonly used acronym for remembering the color and direction is BART—blue away, red toward. The PW Doppler does not continuously transmit and receive the ultrasound pulse. Multiple crystals in the transducer are excited in a quick burst, producing ultrasound waves. This transmission burst is then followed by a "listening" period during which the crystals detect the reflected signals. Signals from more superficial structures are received sooner than those from deeper structures with more extended "listening" periods. This feature allows signals only from specific depths to be processed, thereby controlling sample size and range resolution. Therefore, two vessels located above each other can be evaluated separately, and vessels can also be followed as their courses change. The PW Doppler wave is site-specific and can only measure low-flow velocities—it cannot correctly measure high velocities (above 1.5–1.7 m/s). The CW Doppler uses two piezoelectric crystals, one to emit ultrasound continuously and the other to receive the reflected waves continuously. This results in a fixed sample size and no range resolution or ability to place the sample volume at a specific depth. It also cannot create anatomic images. It is used for precise settings such as very high peak systolic velocities.[1] The CW Doppler measures very high blood flow velocities, and the color flow CF Doppler is a PW Doppler with multiple gates that allow it to measure the flow velocity through the heart on the two-dimensional echocardiographic image. Physicians often use Doppler imaging to detect blockages to the blood flow (due to clots), constriction of vessels or tumors, and congenital vascular malformations. Figure 2-21 shows an ultrasound image of the right common carotid artery and the corresponding Doppler waveform.

Duplex ultrasonography combines physiologic information based on Doppler shift frequencies with anatomic information from real-time, high-resolution B-mode imaging.

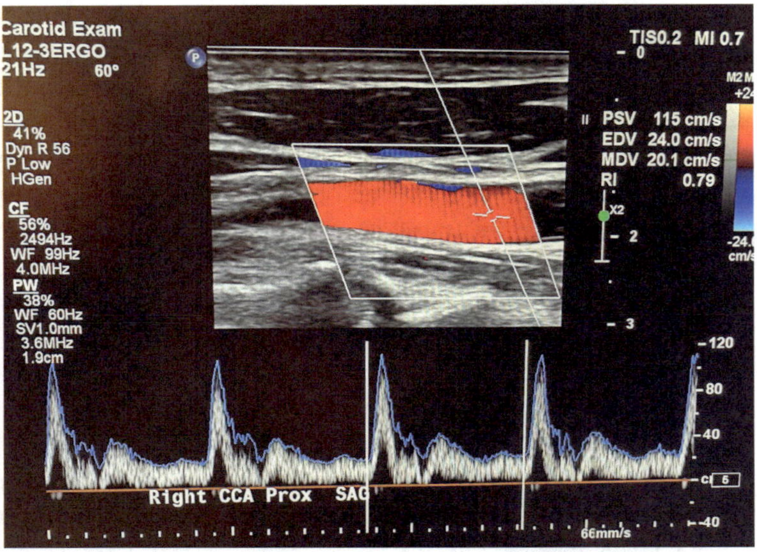

Figure 2-21: Ultrasound image of the right common carotid artery and the corresponding Doppler waveform.

2.13 Blood Flow Dynamics

Understanding blood flow dynamics is important in the study of vascular disease development, such as atherosclerosis, thrombosis, or aneurysms. The circulation system transports nutrients and waste around the body (delivering oxygen and nutrients to the cells and removing cellular wastes and carbon dioxide). Its other function is to maintain a constant temperature and potential or power of hydrogen (pH) in all organs of the body. The circulation system comprises the heart (the pump that drives the blood to all body tissues), blood vessels (delivery routes), and blood (the medium that transports the food and the waste materials). The blood flows continuously through two separate loops that originate and terminate at the heart: the pulmonary circulation and systemic circulation loops, as shown in Figure 2-22. Pulmonary circulation carries blood between the heart and lungs, and systemic circulation carries blood between the heart and the body's organs and tissues. At any time, about 84% of the entire blood volume is in systemic circulation, 7% is in the heart, and 9% is in the pulmonary vessels.

Under normal conditions, the average resting heart rate of an adult between the ages of 18 and 80 is about 75 beats/min, with a stroke volume of 70 mL/beat (cardiac output of 5.25 L/min). For vigorous-intensity physical activity, the heart rate can increase to as high as 200 beats/min, with a stroke volume of up to 150 mL/beat (cardiac output of about 25 L/min).[1] The arteries respond to varying pressure conditions by dilating or shrinking to accommodate the hemodynamic demands.

1. Kisslo J, Adams DB, Belkin RN. Doppler Color Flow Imaging. Baltimore (MD): Churchill Livingstone; 1988. 179 p.

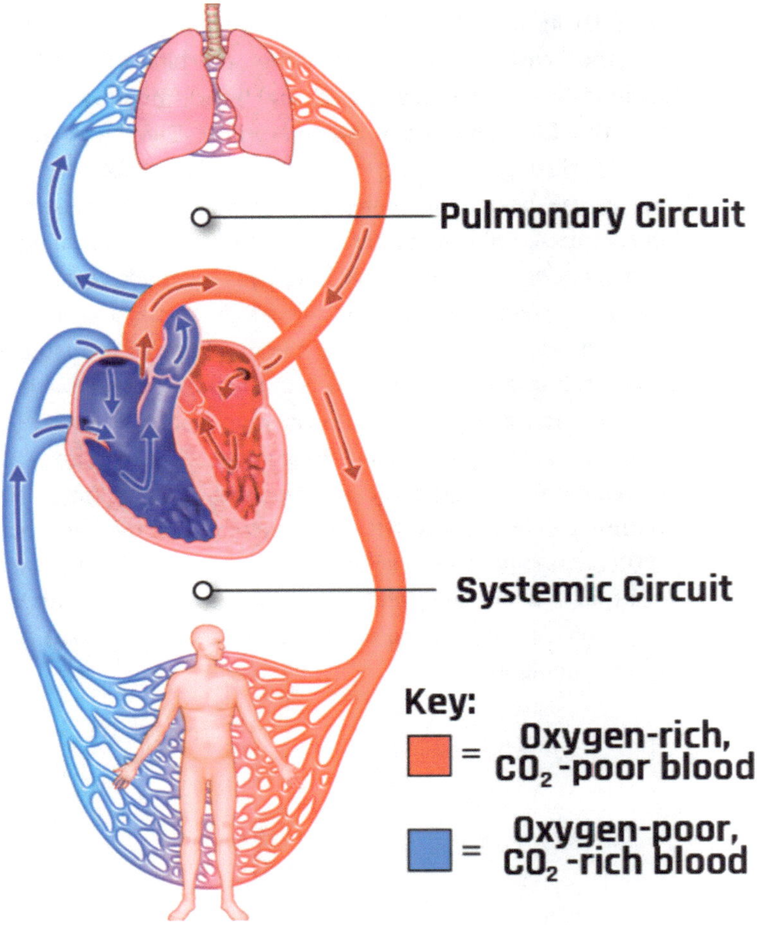

Figure 2-22: Physiological features of the pulmonary and systemic circuits. Diagram of the pulmonary and systemic circuits by Colorado Community College System licensed under CC BY-NC-SA 4.0

The presence of a pressure gradient between the aorta and the veins ensures the blood keeps moving to the peripherals. In mathematical form, the volume per unit time (Q) can be expressed using Dacy's law: $Q = \Delta P/R$, where ΔP is the pressure differential and

R is the resistance. Figure 2-23 shows the systemic blood pressure throughout different paths of the body.

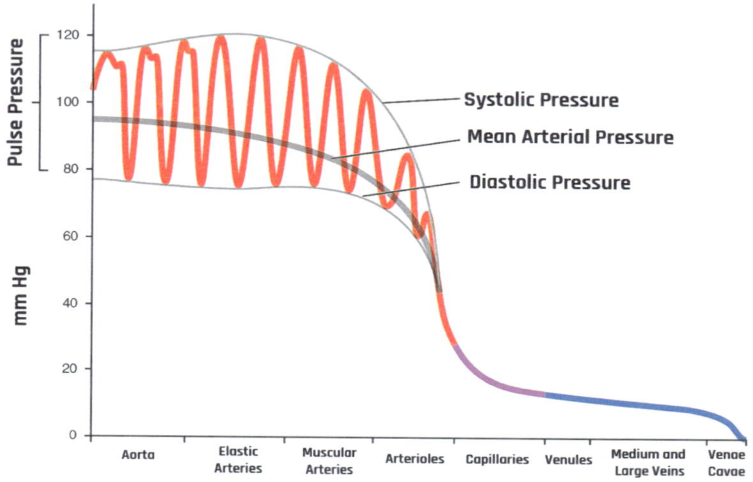

Figure 2-23: Systemic blood pressure throughout different paths of the body. Systemic Blood Pressure by Betts JG, Young KA, Wise JA, Johnson E, Poe B, Kruse DH, Korol O, Johnson JE, Womble M, and DeSaix P. licensed under CC BY 4.0

As the heart pumps the blood, the pressure varies between systolic pressure (pressure peak after ventricular systole) and diastolic pressure (pressure drop during ventricular diastole). In the aorta, the systolic average pressure is about 120 mm of Hg, while the diastolic average pressure is about 80 mm of Hg.

The velocity of the flow is mainly determined by three critical variables: radius (r), vessel length (λ), and viscosity (η), which are related to one another by Poiseuille's equation:

$$\text{Blood flow} = \frac{\Delta P}{\text{Resistance}},$$

where ΔP is the pressure differential and resistance is given by

$$\text{Resistance} = \frac{8\eta\lambda}{\pi r^4}.$$

This shows that since the radius changes with vasoconstriction and vasodilation, the effect will be a dramatic change in the resistance and flow of blood.

The blood flow through straight, long, and smooth vessels is almost linear, with each layer of blood remaining the same distance from the walls of the vessels. These different layers flow at different velocities. Speed is dependent on both the axial distance and pressure. At high pressure, the velocity is high, whereas at low pressure, the velocity is low. This leads to a decrease in pressure and velocity from the heart to peripheral circulation.

Due to the difference in systolic and diastolic pressures, pulse pressure is generated during systole. A pulsatile blood flow is created down the pressure gradient into systemic circulation. The pulse pressure is ~ 40 mm of Hg, the difference between systolic and diastolic pressures.

2.13.1 Laminar Versus Turbulent Flow

In a laminar flow, the motion of the fluid is very orderly, with all particles moving in straight lines parallel to the walls of the tube, as shown on the left of Figure 2-24. The velocity profile across the tube is parabolic, with the fluid's highest velocity at the tube's center, as shown in Figure 2-25. The parabolic profile arises because the fluid molecules touching the walls experience more resistance than those at the center.

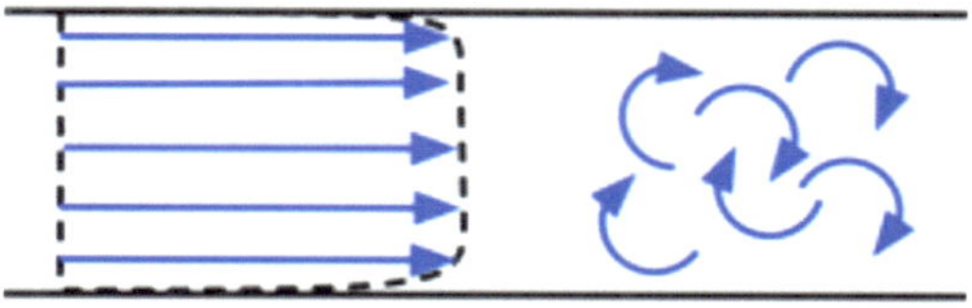

Figure 2-24: Diagram representing laminar flow on the left and turbulent flow on the right. Turbulent pipe flow by Ryan Toomey, University of South Florida licensed under CC BY-SA 4.0

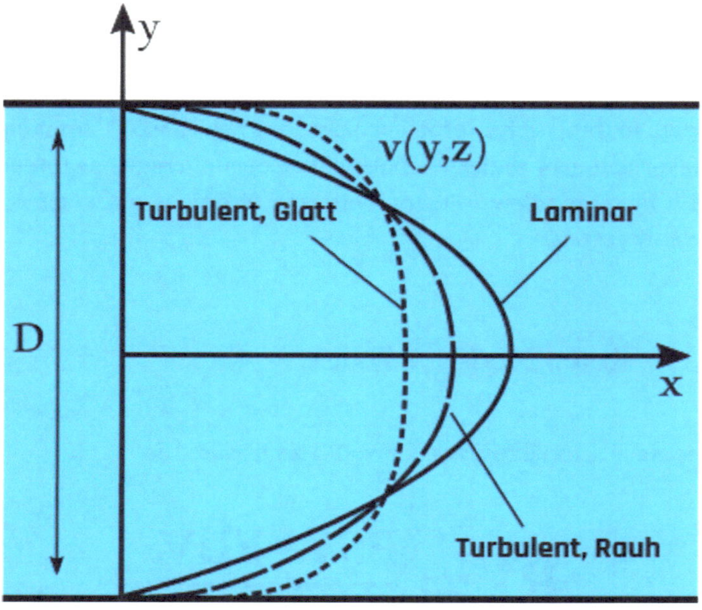

Figure 2-25: Velocity distributions in laminar and turbulent flows. Flow-profile-roughness by Svebert licensed under CC BY-SA 4.0

When the velocity of the blood becomes too high as it passes through a constricted vessel or a rough surface, the flow may become irregular, resulting in random fluctuations in position and time, leading to turbulent flow.

The tendency for turbulent flow is measured using the Reynolds number, which depends on the velocity of the flow, the diameter of the vessel, and the density of the blood:

$$R_e = \frac{vd\rho}{\eta},$$

where ν is the average blood flow velocity (in cm/s), d is the vessel diameter (in cm), ρ is density, and η is the viscosity of the blood. In these units, turbulence occurs when $R_e > 200$, resulting in the formation of eddies. Turbulence can occur in regions of stenosis with increased flow velocity. This type of flow is not common in healthy vessels.

2.13.2 Renal Resistive Index

The renal arterial resistive index (RI) is measured as

$$\mathrm{RI} = \frac{\mathrm{PSV} - \mathrm{EDV}}{\mathrm{PSV}},$$

where PSV and EDV represent the peak systolic velocity and end-diastolic velocity, respectively.

The RI index is used as an indicator for detecting and managing renal artery stenosis, evaluating risk in chronic kidney disease, compiling differential diagnoses in acute and chronic obstructive renal disease, and predicting renal and global outcomes in critically ill patients.

Recent studies have shown that an increased RI reflects changes in intrarenal perfusion and systemic hemodynamics and the presence of subclinical atherosclerosis and, therefore, may provide valuable prognostic information for patients with primary hypertension.[2]

2.14 *Self-Assessment*

1. What is the piezoelectric effect? How does an ultrasound machine work using this effect?
2. What are ultrasound artifacts? What causes them?
3. What is a reverberation artifact? How can you prevent it?
4. Discuss the three display modes in an ultrasound machine.
5. What is the Doppler effect? What is its importance

2. Viazzi F, Leoncini G, Derchi LE, Pontremoli R. Ultrasound Doppler renal resistive index: A useful tool for the management of the hypertensive patient. J Hypertens. 2014 Jan;32(1):149–53. doi: 10.1097/HJH.0b013e328365b29c. PMID: 24172238; PMCID: PMC3868026.

in medical ultrasound imaging?

2.15 *Further Readings*

1. Grogan SP, Mount CA. Ultrasound Physics and Instrumentation. 2023 Mar 27. In: StatPearls [internet]. Treasure Island (FL): StatPearls Publishing; 2023 Jan–. PMID: 34033355.
2. Fischetti AJ, Scott RC. Basic ultrasound beam formation and instrumentation. Clin Tech Small Anim Pract. 2007 Aug;22(3):90–2. doi: 10.1053/j.ctsap.2007.05.002. PMID: 17844814.
3. Quien MM, Saric M. Ultrasound imaging artifacts: How to recognize them and how to avoid them. Echocardiography. 2018 Sep;35(9):1388–1401. doi: 10.1111/echo.14116. Epub 2018 Aug 6. PMID: 30079966.
4. Quiñones MA, Otto CM, Stoddard M, Waggoner A, Zoghbi WA; Doppler Quantification Task Force of the Nomenclature and Standards Committee of the American Society of Echocardiography. Recommendations for quantification of Doppler echocardiography: A report from the Doppler Quantification Task Force of the Nomenclature and Standards Committee of the American Society of Echocardiography. J Am Soc Echocardiogr. 2002 Feb;15(2):167–84. doi: 10.1067/mje.2002.120202. PMID: 11836492.

5. Kisslo J, Adams DB, Belkin RN. Doppler Color Flow Imaging. Baltimore (MD): Churchill Livingstone; 1988. 179 p.

3. 2D, 3D, and 4D Ultrasound Imaging

3.1 Learning Objectives

After reviewing this chapter, you should be able to do the following:

1. Define and explain the mode of operation of two-dimensional (2D), three-dimensional (3D), and four-dimensional (4D) ultrasound imaging.
2. Compare and contrast the advantages and disadvantages of 2D, 3D, and 4D ultrasound imaging in medical applications.
3. Evaluate the diagnostic benefits of 3D/4D ultrasound imaging over 2D imaging in different medical fields, such as obstetrics, gynecology, and cardiology.
4. Analyze the limitations of 3D/4D ultrasound imaging in medical applications, including technical, clinical, and ethical considerations.
5. Describe the recent improvements in technology, image quality, and clinical ease of use that have contributed to the expanding applications of 3D/4D ultrasound imaging in patient care.
6. Critically assess the current use and potential future directions of 3D/4D ultrasound imaging in

clinical practice, research, and education.

7. Discuss the ethical implications of 3D/4D ultrasound imaging, such as fetal imaging for nonmedical purposes, and the impact of imaging on patient autonomy and decision-making.

3.2 Introduction

Medical ultrasound continues to make significant contributions to patient care by providing anatomical information needed by clinicians to make critical decisions. While 3D ultrasound has been commonly used in cardiology, obstetrics, and gynecology, its applications continue to expand as essential technological improvements, image quality, and clinical ease of use increase. In this section, we discuss the modes of operation of 2D, 3D, and 4D ultrasound imaging and their current diagnostic benefits and limitations in the medical field. First, we will consider the advantages of 3D/4D over 2D imaging.

3.3 Using 2D Versus 3D and 4D Ultrasound

Two-dimensional ultrasound is considered a standard or conventional imaging technique. In 2D scanning, a series of thin slices make up an image, and only one slice can be seen at a time.

Three- and four-dimensional clinical ultrasounds have been around for nearly 25 years. However, their use has lagged behind that of computerized tomography (CT) and magnetic resonance imaging (MRI) due to the difficulty in rendering the data in 3D.[1] However, ultrasound equipment's increasing computing power has helped resolve complex signal processing tasks needed to render 3D ultrasound data.

Three-dimensional ultrasound is based on the same principles of operation as 2D ultrasound but has an added position-sensing component to produce the effect of a 3D image, as illustrated in Figure 3-1. In 3D ultrasound imaging, echoes are used to form real-like realistic volume images.

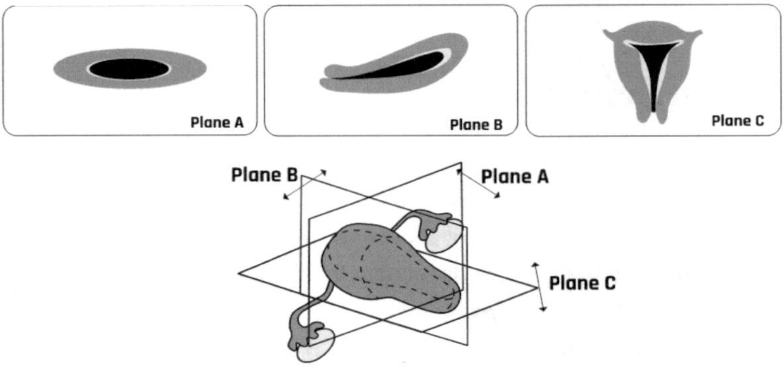

Figure 3-1: Orthogonal planes of the uterus.

A 4D ultrasound shows 3D ultrasound images in motion. Two-dimensional ultrasound images are commonly used because they are less expensive than 3D or 4D. However, many centers now use 3D/4D ultrasound. The most common area of use has been fetal cardiovascular scanning. The 3D/4D technology offers real-time motion-gated cardiac scanning.

A 3D/4D ultrasound offers expecting parents a memorable

1. Benacerraf BR. Three-dimensional fetal sonography: Use and misuse. J Ultrasound Med. 2002 Oct;21(10):1063–7. doi: 10.7863/jum.2002.21.10.1063. PMID: 12369660.

lifetime opportunity to see the features of their unborn babies. Figure 3-2 shows 2D grayscale and 3D colored images.

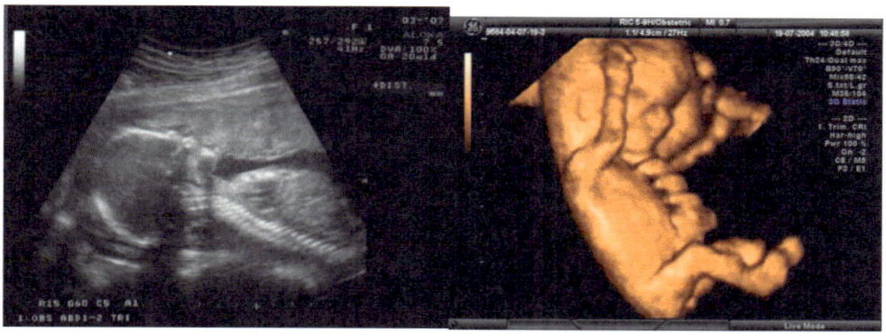

Figure 3-2: 2D grayscale image on the left (Scan20semanas1 by Guimi licensed under CC BY-SA 2.5) and 3D colored image on the right (4215.600×450 by Clayuyu licensed under CC BY-SA 4.0).

3.4 Accuracy and Repeatability

Acquiring qualitative and quantitative sonographic volume data such as multiplanar imaging, surface and volume rendering, and semiautomated volume calculation gives 3D/4D an added advantage over 2D imaging. Virtual planes provide extra information that cannot be viewed with a standard 2D technique.[2] The American Institute of Ultrasound in Medicine (AIUM) notes that the ability to take images in any plane in real time has enormous potential use

2. Yagel S, Cohen SM, Shapiro I, Valsky DV. 3D and 4D ultrasound in fetal cardiac scanning: A new look at the fetal heart. Ultrasound Obstet Gynecol. 2007 Jan;29(1):81–95. doi: 10.1002/uog.3912. PMID: 17200988.

in medical diagnosis.[3] Another advantage is that 3D ultrasound can provide measurements in three planes with acceptable reliability.

3.5 Equipment Design and Image Acquisition

A 3D ultrasound uses a series of 2D images covering a volume of a particular area. This allows the images to be rotated and displayed in different orientations. When displayed in real time, or live, they form a 4D ultrasound. Volume acquisition is achieved by using an array of transducers consisting of many 2D frames, one behind the other. Automated mathematical algorithms are then used to process the volume data to produce the desired image, as illustrated in Figure 3-3.

The reconstruction in Figure 3-3 occurs in a matter of seconds such that the spatio-temporal imaging correlation (STIC) acquisition is completed in the presence of the patient. The STIC acquisition mode can be combined with B-mode, color, and power Doppler. From a good STIC acquisition, a sequential plane may be viewed in corresponding transverse and longitudinal planes at any time point, simultaneously.

3. Benacerraf BR, Benson CB, Abuhamad AZ, Copel JA, Abramowicz JS, Devore GR, Doubilet PM, Lee W, Lev-Toaff AS, Merz E, Nelson TR, O'Neill MJ, Parsons AK, Platt LD, Pretorius DH, Timor-Tritsch IE. Three- and 4-dimensional ultrasound in obstetrics and gynecology: Proceedings of the American Institute of Ultrasound in Medicine Consensus Conference. J Ultrasound Med. 2005 Dec;24(12):1587–97. doi: 10.7863/jum.2005.24.12.1587. PMID: 16301716.

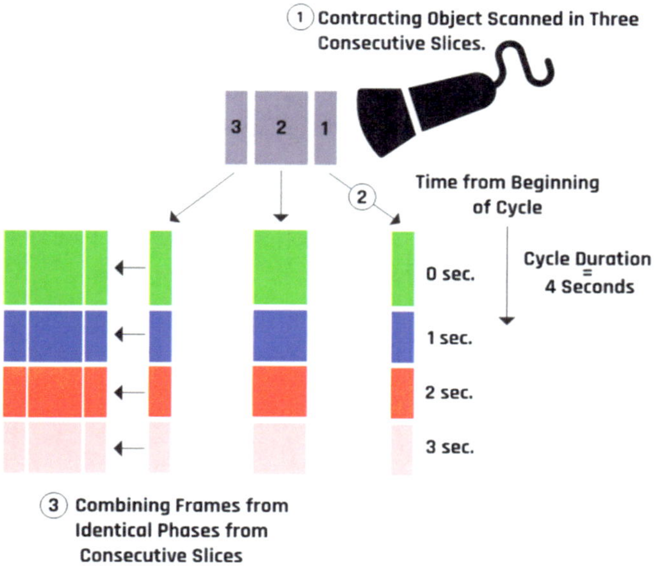

Figure 3-3: A schematic of the formation of a 4D ultrasound image. Figure 1 by Yagel S, Cohen SM, Shapiro I, and Valsky DV licensed under CC BY 2.0

The B-flow image is a live grayscale depiction of blood flow and cardiac chambers. When applied to 3D fetal echocardiography, the B-flow image shows blood flow in the heart and great vessels in real time.[4]

4. Yagel S, Cohen SM, Shapiro I, Valsky DV. 3D and 4D ultrasound in fetal cardiac scanning: A new look at the fetal heart. Ultrasound Obstet Gynecol. 2007 Jan;29(1):81–95. doi: 10.1002/uog.3912. PMID: 17200988.

3.6 Display and Analysis

Four-dimensional imaging involves a 3D image moving in real time, created from a volume of data that allows the live reconstruction of images in different planes and renderings. *Rendering* is taking voxel-based data and converting it into a viewable image with added depth. Three- and four-dimensional imaging function by reconstructing an image from multiple 2D planes.

3.7 Areas of Applications

The 3D color and power Doppler ultrasound technology can be used in vascular abnormality assessments such as the fetal placental cord and pulmonary vessels. Three-dimensional ultrasound is not yet widely used in many routine medical procedures but has been widely used in gynecology, obstetrics, and biopsy.

In gynecology, 3D ultrasound can be used to

- examine and assess the uterus,
- screen for ectopic pregnancies,
- locate intrauterine devices,
- examine the ovaries, and
- perform interventional procedures for patients with infertility.[5]

5. Benacerraf BR, Benson CB, Abuhamad AZ, Copel JA, Abramowicz JS, Devore GR, Doubilet PM, Lee W, Lev-Toaff AS, Merz E, Nelson TR, O'Neill MJ, Parsons AK, Platt LD, Pretorius DH, Timor-Tritsch IE. Three- and 4-dimensional ultrasound in obstetrics and gynecology: Proceedings of the American Institute of Ultrasound in Medicine

In obstetrics, fetal 3D ultrasound can be used to

- examine the nasal bone, ears, and central nervous system;
- evaluate rib and lung volumes;
- screen for spinal cord, placenta, and vertebral abnormalities;
- map abnormalities in multiple gestations; and
- perform fetal cardiac scanning.

The application of 3D/4D ultrasound in biopsy allows for needle tracking in multiple planes simultaneously and imaging of the morphology and proximity of local anesthetic spread around the target nerves.[6] Other specialties in which 3D/4D ultrasound can be applied include dermatology and ophthalmology.

3.8 Advantages of 3D/4D Ultrasound Over 2D Ultrasound

In 2D ultrasound imaging, a series of noncontinuous and presumably representative sections of the imaged organ is used to visualize the anatomy of the organ. Repeated examination is required in order to reconstruct cross sections visually. The absence of quantitative spatial documentation means clinicians must rely on image labels and trust the acquisition technique. Repeated examinations may not produce exact image planes, making

Consensus Conference. J Ultrasound Med. 2005 Dec;24(12):1587–97. doi: 10.7863/jum.2005.24.12.1587. PMID: 16301716.

6. Clendenen NJ, Robards CB, Clendenen SR. A standardized method for 4D ultrasound-guided peripheral nerve blockade and catheter placement. Biomed Res Int. 2014;2014:920538. doi: 10.1155/2014/920538. Epub 2014 Jan 19. PMID: 24575416; PMCID: PMC3915798.

comparing serial exams difficult.[7] In addition, every time an organ is examined, it's likely to give rise to varying measurements of a specific organ or structure, as the imaging process is highly operator-dependent. Volumetric (3D) ultrasound allows volumetric acquisition of anatomic data, which may be needed to examine internal organs such as the liver, gallbladder, gallstones, or kidneys.

In 3D/4D, volume data are obtained in a single image, allowing the operator to view any plane in the volume. The multiplanar rendering mode options also allow for viewing images in real-time motion. Three-dimensional imaging is helpful in examining contours such as those in the facial area, the heart chambers, and the valves. The 4D ultrasound provides a motion video of the 3D structure in real time. Together, 3D and 4D ultrasound have been quite successful in examining the inside or outside of organs, nodules, cysts, or tumors. The 3D ultrasound also offers a more comprehensive image of anatomical structures and pathological conditions. A 3D echocardiography can provide estimates of ventricular volume and function. Comparing 2D and 4D, qualitative 4D ultrasound is considered superior to 2D in real time. However, 4D is inferior to 2D ultrasound for quantitative analysis of movements. For example, 4D evaluation of complex fetal facial activity and expression is better than that of 2D.[8]

Hence 3D and 4D ultrasound imaging allow more accessible and more rapid screening than 2D ultrasound imaging. Irregularly shaped organ volumes can be more accurately visualized using 3D than 2D imaging ultrasound. In addition, 3D measurements are more reproducible than 2D ultrasound and provide the option

7. Fenster A, Downey DB, Cardinal HN. Three-dimensional ultrasound imaging. Phys Med Biol. 2001 May;46(5):R67–99. doi: 10.1088/0031-9155/46/5/201. PMID: 11384074.
8. Benacerraf BR. Three-dimensional fetal sonography: Use and misuse. J Ultrasound Med. 2002 Oct;21(10):1063–7. doi: 10.7863/jum.2002.21.10.1063. PMID: 12369660.

to rotate the volume, which is often necessary to allow optimal visualization of the geometric structure of the organ under examination and determine the position of the structure in the volume.[9]

3.9 Handheld Ultrasound Devices

The market for small, portable ultrasound machines has flourished over the past decade. They have gained popularity and significant usefulness in the medical sector, especially in the emergency medicine department. There are two specific reasons for this popularity: (1) they are smaller/lighter and portable, which makes the whole process of scanning the patients quicker, more accessible, and safer at the same time, and (2) their portability, ease of use, connectivity, and cleanability make them ideal tools for diverse care settings. Conventional large-sized ultrasound machines are demanding and come with certain limits. For instance, these larger machines are laborious and challenging to sterilize thoroughly in comparison to small handheld machines. Also, a physician performing any intervention with this machine usually requires an extra pair of hands to help set it up and obtain clinical images.

Both laptop and handheld models of portable ultrasound systems are available. There are also a few models that are intermediate between these two. Therefore, portable ultrasound machines can be broadly divided into three categories. The first category is the large, laptop-sized device weighing 12 to 15 pounds.

9. Nelson TR. Three-dimensional Ultrasound Imaging. Paper presented at 3D/4D Ultrasound Imaging—UIA Annual Meeting; 2006 Mar 13–14; San Diego, California, USA.

An example of such a device is GE HealthCare's LOGIQ™ *e* portable ultrasound. The second category involves a smaller device, almost the size of a smaller laptop and weighing about 6 to 10 pounds. An example of such a device is the Sonosite Edge machine.

With technological innovations, even smaller and lighter ultrasound devices with better image quality are now available (such as those manufactured by GE HealthCare, Siemens, and Philips). The third category is the handheld ultrasound device that weighs under a pound. Philips's Lumify is an example of one such device; it costs around $6,000 to buy or $2,300 per year to lease. This device operates through a simple USB connection of the transducer to a compatible device such as a smartphone. GE HealthCare's Vscan is a pocket-size ultrasound device that provides real-time gray anatomic and color flow images. It is optimized for physicians to quickly inspect the heart, abdominal organs, and urinary bladder. It can provide insights into areas of obstetrics and gynecology, pleural fluid and motion detection, and pediatrics. A recently extended version of the Vscan handheld ultrasound device can be used to help confirm and monitor the progression of acute respiratory diseases like COVID-19.

3.10 Limitations and Challenges

While 3D/4D ultrasounds show success in some cases, they have some limitations in their clinical applications. Three-dimensional renderings have demonstrated impressive results in some areas; the interfaces are complex and difficult to interpret. Other significant issues raised by the AIUM[10] are that the protocols and manipulation

10. Benacerraf BR, Benson CB, Abuhamad AZ, Copel JA, Abramowicz JS, Devore GR, Doubilet PM, Lee W, Lev-Toaff AS, Merz E, Nelson TR,

techniques are not standardized across manufacturers with respect to the terminology of functions and the display, therefore requiring time and effort to perfect operator skill and accuracy. Standardization is also needed in image orientation. Another challenge is that 3D/4D does not solve poor 2D image problems for several reasons: (1) artifacts and orientation can be confusing, and (2) resolution decreases in reconstructed images such that in some cases, 3D/4D images may be inferior to 2D images. Currently, 2D fetal echocardiography with color Doppler has a success rate of up to 92% in diagnosing congenital heart diseases.[11] Research findings have shown that there are limitations to adequate visualization of fetal anatomy with 3D/4D ultrasound if there is inadequate amniotic fluid surrounding the fetus or if the fetus has its face in the posterior position. Another reported limitation with current 3D/4D Doppler ultrasound technology is that endometrial and subendometrial blood flows measured at one time point during IVF treatment were not good predictors of pregnancy.[12]

The AIUM organized a meeting of physicians and scientists to discuss the diagnostic benefits and technical limitations of 3D ultrasound in obstetrics and gynecology and its potential use in various clinical practices now and in the future. Their

O'Neill MJ, Parsons AK, Platt LD, Pretorius DH, Timor-Tritsch IE. Three- and 4-dimensional ultrasound in obstetrics and gynecology: Proceedings of the American Institute of Ultrasound in Medicine Consensus Conference. J Ultrasound Med. 2005 Dec;24(12):1587–97. doi: 10.7863/jum.2005.24.12.1587. PMID: 16301716.

11. Shen O, Yagel S. The added value of 3D/4D ultrasound imaging in fetal cardiology: Has the promise been fulfilled? Ultrasound Obstet Gynecol. 2010 Mar;35(3):260–2. doi: 10.1002/uog.7569. PMID: 20205202.
12. Benacerraf BR. Three-dimensional fetal sonography: Use and misuse. J Ultrasound Med. 2002 Oct;21(10):1063–7. doi: 10.7863/jum.2002.21.10.1063. PMID: 12369660.

recommendations, together with the equipment manufacturers and other safety regulations, will be reviewed in the next section.

3.11 Self-Assessment

1. What are the fundamental differences between the 2D, 3D, and 4D ultrasounds?
2. How are 3D and color Doppler ultrasound used in gynecology and obstetrics?
3. What are the advantages of 3D and 4D ultrasounds over 2D?
4. What are some of the limitations and challenges of 3D and 4D ultrasounds?

3.12 Further Readings

1. Campbell S. A short history of sonography in obstetrics and gynaecology. Facts Views Vis Obgyn. 2013;5(3):213–29. PMID: 24753947; PMCID: PMC3987368.
2. Gonçalves LF, Lee W, Espinoza J, Romero R. Three- and 4-dimensional ultrasound in obstetric practice: Does it help? J Ultrasound Med. 2005 Dec;24(12):1599–624. doi: 10.7863/jum.2005.24.12.1599. PMID: 16301717; PMCID: PMC7062383.

3. Yagel S, Cohen SM, Shapiro I, Valsky DV. 3D and 4D ultrasound in fetal cardiac scanning: A new look at the fetal heart. Ultrasound Obstet Gynecol. 2007 Jan;29(1):81–95. doi: 10.1002/uog.3912. PMID: 17200988.
4. Fenster A, Downey DB. Three-dimensional ultrasound imaging. Annu Rev Biomed Eng. 2000;2:457–75. doi: 10.1146/annurev.bioeng.2.1.457. PMID: 11701520.

4. Ultrasound Bioeffects and Safety

4.1 Learning Objectives

After reviewing this chapter, you should be able to do the following:

1. Define the concept of ultrasound bioeffects and describe the various types of bioeffects that can occur during medical ultrasound procedures.
2. Explain the physical mechanisms that underlie ultrasound bioeffects, including thermal and nonthermal effects.
3. Describe the factors that can influence the likelihood and severity of ultrasound bioeffects, such as the frequency, intensity, and duration of ultrasound exposure.
4. Discuss the potential clinical consequences of ultrasound bioeffects, including tissue damage, organ dysfunction, and other adverse outcomes.
5. Outline the current safety guidelines and regulations related to medical ultrasound use, including those established by professional organizations and regulatory agencies.
6. Identify the key safety considerations and best practices for using ultrasound in clinical practice,

including proper equipment calibration, patient positioning, and operator training.

4.2 Introduction

While no medical procedure is entirely risk-free, ultrasound is considered the safest imaging method currently in use. This is because no magnetic field or radiation is used in the imaging process. However, the absence of evidence of harm from any medical procedure does not provide proof of the absence of risks. Therefore, clinicians must take precautions when using any medical technique.

When using ultrasound, risks must be minimized to ensure patient safety. There are two basic rules of medical ethics:[1]

- *primum non nocere* (translated as "first, do no harm"), and
- perform a procedure only if the importance of doing so outweighs the risk.

In that case, a procedure is deemed ethical only when it benefits the patient.

Ultrasound is generally considered safe if used properly. While there is no evidence it will physically affect the patient long-term, it should still be noted that it is a form of energy, and even

1. Ter Haar G. Ultrasonic imaging: Safety considerations. Interface Focus. 2011 Aug 6;1(4):686–97. doi: 10.1098/rsfs.2011.0029. Epub 2011 May 25. PMID: 22866238; PMCID: PMC3262273.

at low levels, some studies suggest that overexposure to ultrasound can lead to potential undesirable effects such as

- heat damage to tissue due to energy absorption;
- hearing damage caused by high-frequency sounds from the ultrasound machine; and
- nausea, headaches, and fatigue.

The damage type and extent mainly depend on the ultrasound wave characteristics such as frequency, intensity level, and exposure time, among other factors.

4.3 Acoustic Output Labeling

There are two possible biological effects of ultrasound exposure: (1) thermal (heating) effect and (2) mechanical (cavitation and streaming) effect. These effects can be quantified by reading the mechanical index (MI) and the thermal index (TI) on the ultrasound output display. The MI and TI are on-screen indicators of the potential bioeffects of ultrasound exposure. While imperfect, the TI and MI are considered good thermal and nonthermal risk assessments. These are discussed in detail in the sections below.

4.3.1 Thermal Index

As the ultrasound traverses the tissue, it absorbs some of the energy, causing the tissue temperature to increase. The rate at which the energy is absorbed depends on

- the time of exposure,
- the intensity of the ultrasound beam,

- the attenuation effect of the tissue, and
- the frequency of the ultrasound beam.

Higher thermal effects occur in tissues with higher absorption coefficients (such as bones), and lower thermal effects occur in tissues with lower absorption coefficients (such as amniotic fluids).[2] There is a concern that the embryo and fetus are particularly susceptible to heat energy.

The TI is a relative indicator of thermal risk for the likely temperature rise that might be produced after prolonged exposure. A more significant TI value represents a higher temperature and a higher risk. Three kinds of TI values can be displayed on an ultrasound machine depending on the type of application:[3]

- The TI for soft tissues (TIS) assumes that the ultrasound beam does not impinge on bone, such as is the case for the first trimester.
- The TI for bones (TIB) assumes that the beam impinges on bone at or near its focus, such as is the case for the second and third trimesters.
- The TI for cranial bone (TIC) assumes that the transducer front face is very close to the bone, such as is the case when scanning the adult cranium.

The thermal index (TI) is calculated as the ratio of the acoustic power produced by the transducer (Wp) to the power

2. Gibbs V, Cole D, Sassano A. Ultrasound Physics and Technology: How, Why and When? [place unknown]: Elsevier Health Sciences; 2009. 144 p.
3. Nelson TR, Fowlkes JB, Abramowicz JS, Church CC. Ultrasound biosafety considerations for the practicing sonographer and sonologist. J Ultrasound Med. 2009 Feb;28(2):139–50. doi: 10.7863/jum.2009.28.2.139. PMID: 19168764.

required by the tissue to raise its temperature by 1 degree Celsius (W_{Deg}). In mathematical form:

$$TP = \frac{W_p}{W_{Deg}}.$$

The index allows sonographers and clinicians to assess the relative potential heating effects associated with ultrasound imaging. A thermal index of 1 indicates the amount of acoustic power required to raise the tissue temperature by 1 degree Celsius. A higher TI value indicates a higher temperature and a higher risk.

In general, tissue heating by ultrasound is related directly to the intensity of ultrasound waves. The rate of increase in temperature is related to the ultrasound intensity and degree of absorption. It is inversely proportional to tissue density and specific heat. The rate at which the tissue absorbs the heat depends on the protein concentration. For example, fat heats much faster than dense tissue such as muscles.

Nevertheless, the tissue temperature rise is also limited by the cooling effects of the blood flow, which makes it more challenging to heat vascular organs (such as the liver and kidney) than bone.[4] Another concern is that the presence of bones increases the likelihood of a temperature rise due to absorption in the bone and the conduction of heat from bone to adjacent tissues. However, TI values of greater than 1.5 can occur on Doppler ultrasound.

4. Ter Haar G. Ultrasonic imaging: Safety considerations. Interface Focus. 2011 Aug 6;1(4):686–97. doi: 10.1098/rsfs.2011.0029. Epub 2011 May 25. PMID: 22866238; PMCID: PMC3262273.

Therefore, prolonged pulsed Doppler ultrasound is not recommended for sensitive tissues such as those of the embryo (less than eight weeks), eye, head, brain, and spine.[5]

4.3.2 Mechanical Index

Oscillations of gas bubbles (cavitation) occur inside tissue due to ultrasound pressure waves. The cavitation effect happens due to the excitation of a stable gas bubble by an acoustic field (noninertial cavitation—the formed bubble oscillates in the acoustic field), and streaming effects result from the movement of complex fluids due to radiation force pressures (inertial, or transient, cavitation—the formed bubble rapidly collapses and produces a shock wave that can be capable of causing biological damage).[6, 7] Bubbles are formed in a liquid when the local pressure falls (rarefaction part of the ultrasound wave) below the vapor pressure of the liquid. This cavitation effect is dependent on the fluid inertia, viscosity, and surface tension. Therefore, it is important to shorten the exposure

5. Nelson TR, Fowlkes JB, Abramowicz JS, Church CC. Ultrasound biosafety considerations for the practicing sonographer and sonologist. J Ultrasound Med. 2009 Feb;28(2):139–50. doi: 10.7863/jum.2009.28.2.139. PMID: 19168764.
6. Nelson TR, Fowlkes JB, Abramowicz JS, Church CC. Ultrasound biosafety considerations for the practicing sonographer and sonologist. J Ultrasound Med. 2009 Feb;28(2):139–50. doi: 10.7863/jum.2009.28.2.139. PMID: 19168764.
7. Bigelow TA, Church CC, Sandstrom K, Abbott JG, Ziskin MC, Edmonds PD, Herman B, Thomenius KE, Teo TJ. The thermal index: Its strengths, weaknesses, and proposed improvements. J Ultrasound Med. 2011 May;30(5):714–34. doi: 10.7863/jum.2011.30.5.714. PMID: 21527623.

time to minimize cavitation. Another critical point is that cavitation is likely to occur at lower transducer frequencies. However, cavitation-related bioeffects require the presence of cavitation nuclei (or bubbles) close to cells and physical or chemical interaction between bubbles and cells.

The use of the MI as an indicator is based on the assumption that sound induces oscillations of microbubbles, which can cause an increase in the internal temperature of a water gas bubble.

Nevertheless, tissue viscosity is 100 times greater than that of water such that bubble oscillations are greatly limited. A typical display of soft tissue TI and MI values for a carotid exam is shown in the top right corner of the Doppler image in Figure 4-1.

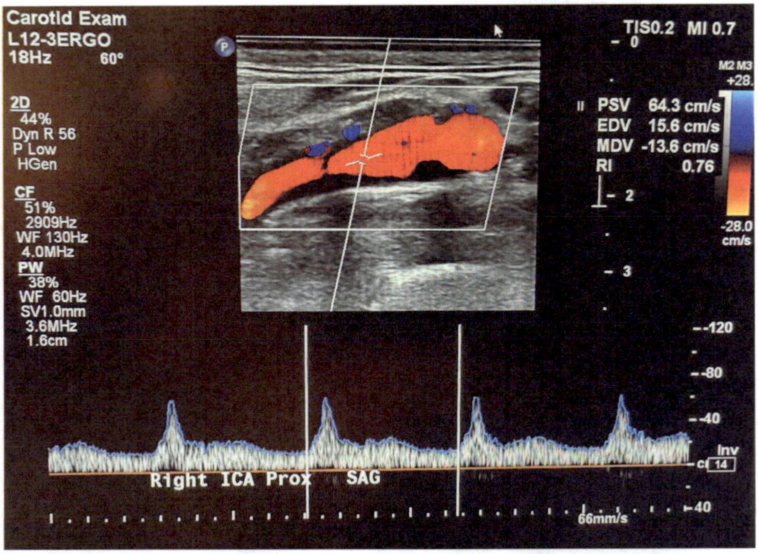

Figure 4-1: Ultrasound image showing the thermal index and mechanical index of a carotid exam in the top right corner.

The risk of cavitation increases with increasing MI values. Other studies suggest that high MI values are associated with the induction of premature ventricular contractions in echocardiography.[3] The potential for nonthermal biohazards is likely to increase if the equipment is not used correctly. Such biohazards have been observed in animal tissues with gas bubbles at MI values greater than 0.3.[83] The bioeffects associated with MI values of 1.9 or less have been reported in skeletal muscle, fat, myocardia, kidneys, livers, and intestines.[9]

Consequently, unnecessary exposure to tissues such as the neonatal lungs should be avoided. Ultrasound operators should keep the MI values as low as possible when carrying out a diagnosis.

4.3.3 Other Indicators

As discussed earlier, the other indicators of thermal effects include the TIS, TIB, and TIC values. The TIS is used when the ultrasound beam passes through soft tissue only, such as in the abdomen and fetal examination during the first trimester. If bones are present, the TIC is used (e.g., in head examinations). The TIB is used if the ultrasound encounters a bone (e.g., in the fetal examination in the second and third trimesters). In addition, the TIS is commonly used for fetal scanning during the first eight weeks of gestation, and the TIB is used after eight weeks. In adults, the TIS is commonly used for

8. Nelson TR, Fowlkes JB, Abramowicz JS, Church CC. Ultrasound biosafety considerations for the practicing sonographer and sonologist. J Ultrasound Med. 2009 Feb;28(2):139–50. doi: 10.7863/jum.2009.28.2.139. PMID: 19168764.
9. Miller DL, Averkiou MA, Brayman AA, Everbach EC, Holland CK, Wible JH Jr, Wu J. Bioeffects considerations for diagnostic ultrasound contrast agents. J Ultrasound Med. 2008 Apr;27(4):611–32; quiz 633–6. doi: 10.7863/jum.2008.27.4.611. PMID: 18359911.

eye scanning. In pediatric and adult patients, the TIC is used when examining areas close to skull bone. The TIB is used for all other purposes.

4.4 Potential Electrical and Mechanical Hazards

Following the correct electrical safety procedures reduces the possibility of fire, muscle contracture, and tissue burns. Shocks can occur when a person touches electrical wires due to broken insulation in the circuit, which may contract muscles, alter brain function, or lead to ventricular fibrillation. Small currents can harm patients through saline catheters, which convert them to relatively large current densities. To reduce patient susceptibility to electrical hazards, equipment with any broken or uninsulated cables and damaged probes must not be used.

4.5 Epidemiology

A major concern in the safety of ultrasound has been the exposure of the embryo and fetus during pregnancy examination. Based on current knowledge, there is insufficient information to link diagnostic ultrasound and recognized adverse effects in humans.

4.6 Guidelines and Regulations

The use of diagnostic ultrasound equipment is limited to medical diagnosis only. It can only be used by personnel who are fully trained in the safe and proper operation of the equipment. Other expectations are that the operator should have a complete awareness of machine settings and understand the effect of those machine settings on thermal and mechanical bioeffects. The simple rule that must be followed is that the default power setting protocols should be set at "**A**s **L**ow **A**s (is) **R**easonably **A**chievable" (**ALARA**) to produce diagnostic quality images. In order to minimize the exposure time on a specific anatomical structure, it is recommended to move or lift the probe when stationary imaging is not necessary. When feasible, it is advisable to avoid imaging fields of view that contain sensitive tissues, including the eye, lungs, and intestines (gas-filled tissues) and fetal calcified structures, such as the skull and spine.

The AIUM, the U.S. Food and Drug Administration (U.S. FDA), and other organizations have issued some guidelines on the use of medical ultrasound in the United States. These guidelines are presented in the sections below.

4.6.1 American Institute of Ultrasound in Medicine Guidelines

The AIUM has issued minimum criteria for a complete medical examination of different parts of the body.[10] According to the AIUM guidelines, there are no independently confirmed significant biological effects in the low megahertz frequency range for ultrasound with intensities below 100 mW/cm^2. In addition, for focused ultrasound, such effects have not been proven even at

10. American Institute of Ultrasound in Medicine. Medical Ultrasound Safety. 3rd ed. Laurel, MD: American Institute of Ultrasound in Medicine; 2014.

higher intensities for exposure times below 500 s. When adjusting controls that influence acoustic output, the AIUM stresses that it is essential to adhere to the ALARA principle and take into account both the duration of transducer dwell and the total scanning time. It is important to be knowledgeable about the MI upper limit, TI upper limit, and associated duration limits for the examination type being conducted. For example, the guidelines established by Harris et al.[11] and endorsed by the AIUM for the recommended maximum duration of ultrasound exposure at a given setting of the TI are presented in Tables 4-1 and 4-2.

11. Harris GR, Church CC, Dalecki D, Ziskin MC, Bagley JE; American Institute of Ultrasound in Medicine; Health Canada; British Medical Ultrasound Society. Comparison of Thermal Safety Practice Guidelines for Diagnostic Ultrasound Exposures. Ultrasound Med Biol. 2016 Feb;42(2):345–57. doi: 10.1016/j.ultrasmedbio.2015.09.016. Epub 2015 Nov 28. PMID: 26626492.

Table 4-1: Recommended maximum exposure time and TI ranges for obstetric (including gynecologic when pregnancy is possible), neonatal transcranial, and neonatal spinal examinations.

TI range	Time (min)
>3.0	0
2.5–3.0	<1
2.0–2.5	<4
1.5–2.0	<15
1.0–1.5	<30
0.7–1.0	<60
<0.7	No limit

Table 4-2: Recommended maximum exposure time and TI ranges for adult transcranial, general abdominal, peripheral vascular, neonatal (except head and spine), and other scanning examinations (except the eye).

TI range	Time (min)
>6.0	0
5.0–6.0	<0.25 (15 s)
4.0–5.0	<1
3.0–4.0	<4
2.5–3.0	<15
2.0–2.5	<60
1.5–2.0	<120
<1.5	No limit

Even if it is necessary to go beyond the recommended values, an undesirable thermal effect is unlikely to occur in most scanning environments due to attenuating factors such as sensor motion and tissue perfusion. However, it is important to adhere to the ALARA principle so that the duration of the examination is limited to the amount of time needed to obtain a useful diagnostic result.

4.6.2 U.S. Food and Drug Administration Guidelines

Because of the likelihood that ultrasound may have some potential heating and mechanical effects, the U.S. FDA states that the use of ultrasound is not entirely harmless. For this reason, it discourages the nonmedical use of ultrasound devices. The guideline established by the U.S. FDA uses the spatial peak pulse-average intensity (I_{sppa}) and the spatial peak temporal-average intensity (I_{spta}) in addition to the MI as safety measures for diagnostic ultrasound.[12] I_{spta} indicates the highest intensity measured at any point in the ultrasound beam averaged over the temporal (time) duration of the pulse, and I_{sppa} represents the highest intensity measured at any point in the ultrasound beam averaged over the pulse repetition period. Table 4-3 lists the highest limit values for these acoustic indicators for diagnostic ultrasound devices.

Table 4-3: U.S. FDA-issued acoustic output exposure level limits.

Use	I_{spta} (mW/cm^2)	I_{sppa} (W/cm^2) or MI
Peripheral vessel	720	190 1.9
Cardiac	430	190 1.9
Fetal imaging and others	94	190 1.9
Ophthalmic	17	28 0.23

Other categories that could be included in Table 4-3 are abdominal, intraoperative, pediatric, and small organ (breast, thyroid, testes, etc.). The U.S. FDA also mandates the display of the likelihood of ultrasound-induced bioeffects, known as the Standard for Real-Time Display of Thermal and Mechanical Acoustic Output

12. United States Department of Health and Human Services, United States Food and Drug Administration, United States Center for Devices and Radiological Health. Marketing clearance of diagnostic ultrasound systems and transducers: Guidance for industry and Food and Drug Administration staff. Silver Spring, MD: Center for Devices and Radiological Health; 2019 Jun 27. 60 p.

Indices, on diagnostic ultrasound equipment capable of producing higher thermal or mechanical effects. The U.S. FDA discourages any unapproved use of medical devices without a physician's order. Such practices may be in violation of state laws or regulations. The operator has the ultimate responsibility for the safe use of ultrasound equipment.

4.6.3 Other Professional Societies/Organizations' Guidelines

The National Collaborating Centre for Women's and Children's Health recommends fetal ultrasound screening between 18 and 20 weeks gestation.[13] The guideline was issued on the basis that there is no evidence to support the need for routine use of ultrasound screening after 24 weeks gestation. The Society of Maternal and Fetal Medicine[14] recommended only one medically indicated ultrasound per pregnancy.

4.7 Hygienic Considerations

Medical equipment, including ultrasound machines and

13. National Collaborating Centre for Women's and Children's Health (UK). Antenatal Care: Routine Care for the Healthy Pregnant Woman. London: RCOG Press; 2008 Mar. PMID: 21370514.
14. Society for Maternal-Fetal Medicine (SMFM), Coding Committee. White paper on ultrasound code 76811. Washington, DC: SMFM; 2004 May 24. Available from: http://www.smfm.org/index.cfm?zone=news&nav=viewnews&newsID=238&smfmon=yes

transducers, may act as both sources and vectors of microbial transmission during examination. Globally, medical equipment-acquired infections are on the increase. Several studies have confirmed the transmission of bacteria and viruses through improper hygienic use of medical equipment. Common transmittable bacteria and viruses that have been identified include *Staphylococcus aureus*, *Pseudomonas*, *Acinetobacter* species, *Candida albicans*, hepatitis B, hepatitis C, human immunodeficiency virus, and herpes.[15]

Just as with any medical procedure, there is a potential risk for cross infection (patient-to-patient, patient-to-operator, or operator-to-patient) if proper procedures are not followed. The risk of infection is highest in procedures that utilize transducers in intracavities or where body fluids are encountered. These risks are classified into three categories:

- **Critical:** Equipment used in settings with a high risk for infection must be sterilized at the time of use. This includes any objects that enter tissue or have been exposed to body fluids.
- **Semicritical:** Equipment or objects used in procedures in which the items encounter body fluids or nonintact skin should be used once or thoroughly disinfected after every use.
- **Noncritical:** Equipment that comes in contact with intact skin must be used once and disposed of or sufficiently disinfected after every use.

The cleaning, disinfecting, and sterilizing of reusable medical and surgical equipment or devices must comply with the manufacturer's requirements and safety protocols. Routine cleaning

15. Sahu B, Raine-Fenning N. Ultrasound and the risk of nosocomial cross infection. Ultrasound Obstet Gynecol. 2010 Aug;36(2):131–3. doi: 10.1002/uog.7729. PMID: 20681005.

must include removing the coupling gel and any visible residue from the probe using the recommended detergent. Disposable gloves must be worn during the cleaning process. Areas that require extensive cleaning or disinfection are the transducer probes and ultrasound machine screens. The keyboard is another critical area that can be easily forgotten during cleaning or disinfection. The choice of the keyboard matters—keyboards similar to those used for computers are not ideal for medical purposes, as it is impossible to clean in between the keys after every medical examination. Most modern ultrasound keyboards now come with disinfectable polyurethane covers, which make them easier to clean.

A simple rule is to regard every patient as a potential source of infection. To prevent the risk of cross infection, the following minimum precautions are recommended in every examination:

- Maintain hand hygiene before and after each examination.
- Use personal protective equipment.
- Use clean and disinfected equipment.
- Maintain a clean working environment.
- Correctly dispose of waste.

4.8 Bioethics

While the use of ultrasound in medicine has gained traction over the past two decades, it still faces significant hurdles in some parts of the globe. This is partly due to differences in culture. For example, in some countries, ultrasound is now used for sex-selective abortion.[16] The purpose of this book is not to debate these ethical

16. Tong Y. Changes in son preference, ultrasound use and fertility [internet]. PEW RESEARCH CENTER; 2022 Aug 23. Available from: https://www.ncbi.nlm.nih.gov/books/NBK7274/

issues but to bring an awareness of how different societies view the use of ultrasound in medicine.

Clinicians often find themselves in difficult situations, especially when diagnosed anomalies require critical procedures. For example, some fetal abnormalities present ethical dilemmas concerning patient counseling, and abortion issues present challenges in gynecology and obstetrics.

That means clinicians have to play a pivotal role in educating patients about the need to perform these procedures and their benefits to allay the emotional nature of the ultrasound diagnosis or obstetric scanning and the associated implications. In addition, clinicians performing the screening may face dilemmas when diagnosing life-threatening problems with no corrective solutions. In some sections of society, revealing the life span of a patient may not be well received or accepted. That leaves the burden of counseling to the clinicians.

4.9 Self-Assessment

1. Why is ultrasound considered relatively safe compared to other imaging equipment?
2. What are the two possible biological effects of ultrasound exposure? Briefly discuss them.
3. What is the ALARA protocol?
4. What are some of the hygienic considerations to remember during and after using ultrasound?

4.10 Further Readings

1. American Institute of Ultrasound in Medicine. Recommended maximum scanning times for displayed thermal index (TI) values. Laurel, MD: American Institute of Ultrasound in Medicine website; 2016 Oct 30. Available from: https://www.aium.org/resources/statements.aspx
2. American Institute of Ultrasound in Medicine. Medical Ultrasound Safety. 3rd ed. Laurel, MD: American Institute of Ultrasound in Medicine; 2014.
3. Quarato CMI, Lacedonia D, Salvemini M, Tuccari G, Mastrodonato G, Villani R, Fiore LA, Scioscia G, Mirijello A, Saponara A, Sperandeo M. A Review on Biological Effects of Ultrasounds: Key Messages for Clinicians. Diagnostics (Basel). 2023 Feb 23;13(5):855. doi: 10.3390/diagnostics13050855. PMID: 36899998; PMCID: PMC10001275.

5. Obstetric Ultrasound

5.1 Learning Objectives

After reviewing this chapter, you should be able to do the following:

1. Understand the embryologic development in the first, second, and third trimesters.
2. Examine various fetal structures and identify abnormalities.

5.2 Introduction

The exciting notion that an individual organism has embryologic development that reflects the species' evolutionary development has been a theory that is present not only in science but also in other fields such as language and music. With the use of ultrasound, we will be able to observe the incredible development of a fertilized ovum into an infant. An example of this transition has been shown in Figure 5-1.

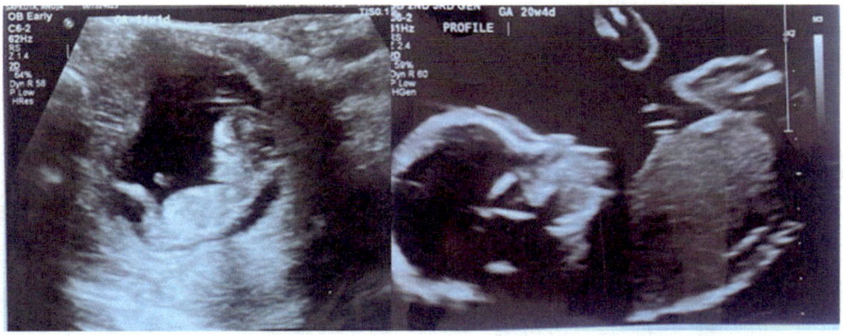

Figure 5-1: Fetal development from the 9th week to the 20th week.

5.3 Role of Ultrasound in Obstetrics

Obstetrics encompasses some of the same considerations as adult diagnostic ultrasound with a critical addition: rapid growth. There is no other time in life when various organs' functionality is time dependent. The developmental changes in early organ growth are part of the most fascinating differentiation process. What is being observed within a specific imaging procedure in utero, although completely normal, actually changes over time. Broken down further, findings at specific gestational times depend on the level of differentiation an organ has undergone at a specific time in gestational development.

The accuracy of ultrasound is variable, and the issue of diagnostic accuracy is essential to understand for any medical professional. *Sensitivity* and *specificity* are commonly used quantitative measures to report such accuracy. For a given test and disease/condition, *sensitivity* is the ability of the test to correctly identify something being positive. Its *specificity* for a given test and disease/condition is how well it can distinguish those with disease from those without.[1] Expressed as equations, the calculations may be done as follows:

$$\text{Specificity} = \frac{\text{TP}}{\text{TP} + \text{FN}}$$

and

$$\text{Sensitivity} = \frac{\text{TN}}{\text{TN} + \text{FP}} ,$$

where

- true positive (TP) results when an imaging test is positive, and the patient has the disease/condition;
- false positive (FP) results when an imaging test is positive, and the patient does not have the disease/condition;
- true negative (TN) results when an imaging test is negative, and the patient does not have the disease/condition; and
- false negative (FN) results when an imaging test is negative, and the patient has the disease/condition.

In recent years more than ever, another consideration is the cost of medical interventions. Clinicians should have a good idea of the impact of the tests that they order:

1. Morgan M, Sciacca F, Knipe H. Sensitivity and specificity. Radiopaedia [internet]. [date unknown] [cited 2023 Oct 28]. Available from: https://doi.org/10.53347/rID-34845.

1. How much does a test cost the patient and society?
2. What is the sensitivity and specificity of each test?
3. Will the test results aid in the care of the patient?

An example of a standard test is the complete blood count (CBC). The CBC is relatively inexpensive, reproducible among different labs, and often clinically helpful. The sensitivity of the CBC is regarded as relatively high. The specificity is often not very high. On this latter point, an elevated CBC can often be one of the first and most important clues to an impending worsening of an infectious process such as appendicitis. An elevated CBC can also be considered a "false positive" caused by medicines such as steroids or endogenous epinephrine secretion (for example, if the patient recently experienced a traumatic event). In the early stages of pregnancy, a CBC test can help determine various health abnormalities, including anemia and infections.

Maternal-fetal medicine specialists and their specialized technicians understand fetal development and testing rationales to make truly unique diagnoses. It is not unusual for heart development to be assessed weekly or even more frequently when difficulties are suspected, as cardiac problems can develop as gestation progresses. The following basic assessments are commonly performed in obstetrics ultrasound.

5.3.1 Amniotic Fluid Assessment

The volume of amniotic fluid is calculated by what is called the *amniotic fluid index*. Even debris (represented by diffuse abnormal reflections in the amniotic fluid) is a harbinger of difficulties. At first glance, the ease of clarity and measurement is heavily influenced by the amount of amniotic fluid. As discussed in the earlier chapters, media, especially fluid, influences the quality of ultrasound wave transmission and reflection. Figure 5-2 shows an ultrasound image at 23 weeks of the fetus, amniotic fluid, and normal fetal morphology. A clear initial view is not necessarily a good sign, since excess amniotic fluid (a condition referred to as *polyhydramnios*) may make the images easy to obtain but may, more importantly, indicate problems. It has been noted that polyhydramnios represents a high-risk obstetric condition as much as 20% of the time.[2,3]

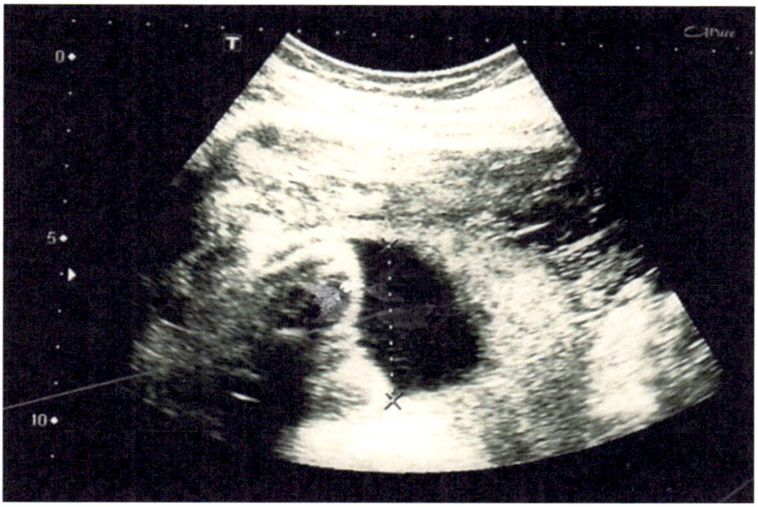

Figure 5-2: Ultrasound image at 23 weeks showing fetus, amniotic fluid, and normal fetal morphology. Ultrasonography picture at 23 weeks showing fetus, amniotic fluid and normal fetal morphology by Dahab AA, *Aburass* R, *Shawkat* W, *Babgi* R, *Essa* O, *and Mujallid* RH *licensed under* CC BY 2.0

2. Hwang DS, Mahdy H. Polyhydramnios. [Updated 2023 Feb 20]. In: StatPearls [internet]. Treasure Island (FL): StatPearls Publishing; 2023 Jan–. Available from: https://www.ncbi.nlm.nih.gov/books/NBK562140/
3. Hamza A, Herr D, Solomayer EF, Meyberg-Solomayer G. Polyhydramnios: Causes, Diagnosis and Therapy. Geburtshilfe Frauenheilkd. 2013 Dec;73(12):1241–1246. doi: 10.1055/s-0033-1360163. PMID: 24771905; PMCID: PMC3964358.

5.3.2 Heart Rate and Heart Development

Fetal cardiac development (the development of the four chambers) is fascinating. The normal opening between the two upper chambers of the fetus's heart, the right and left atria, is called the foramen ovale (FO). The FO permits blood flow to bypass the lungs before the infant is born (a fetus gets its oxygen from the placenta, not the lungs). As a result, the heart does not have to work as hard to pump blood where it is not needed. The FO typically closes six months to a year after the infant is born. A patent foramen ovale (PFO) occurs when the FO remains open after birth. A PFO often does not cause any issues.[4] When an infant is born with congenital heart abnormalities, the FO is more likely to remain open.

There is a division of a single atrium and a single ventricle by the growth of a septal wall from the wall of the single cavity in all directions toward the middle. This is a gradual and persistent hole in the septum that grows abnormally, leading to the septum closing, which results in a septal defect. Fortunately, one can detect both an atrial septal defect and/or a ventricular septal defect using ultrasound, as shown in Figure 5-3. Color flow ultrasound Doppler is often helpful, as one can view the flow from one side of the dividing atrial atrium or ventricle to the other side.

4. Hampton T, Alsaleem M, Murphy-Lavoie HM. Patent Foramen Ovale. [Updated 2022 Sep 12]. In: StatPearls [internet]. Treasure Island (FL): StatPearls Publishing; 2023 Jan–. Available from: https://www.ncbi.nlm.nih.gov/books/NBK493151/

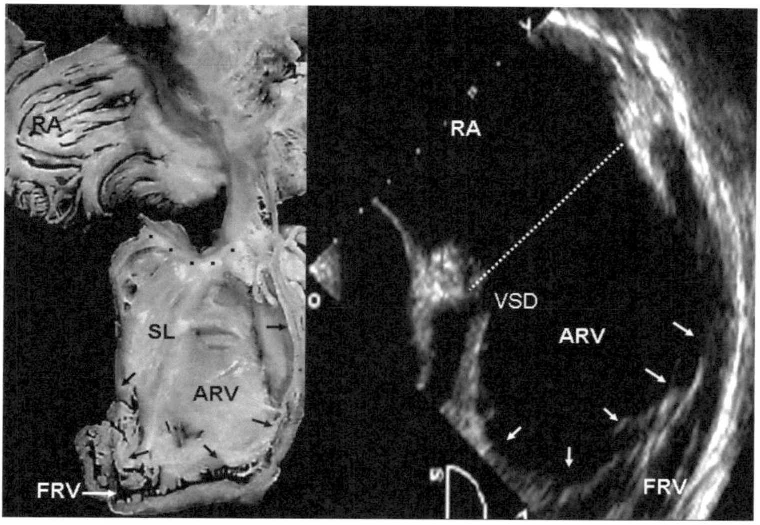

Figure 5-3: The figure on the left shows the internal view of the cardiac right chambers, and the one on the right is the echocardiographic image showing the ventricular septal defect. Internal view of cardiac right chambers by Muñoz-Castellanos L, Espinola-Zavaleta N, Kuri-Nivón M, and Keirns C. licensed under CC BY 2.0

5.3.3 The Relationship Between the Maturity of Different Organs

Different gestational sizes of different organs in the same fetus can indicate genetic and fetal nutritional problems. Intrauterine Growth Retardation (IUGR) can be either symmetric or asymmetric:

(A) Symmetric indicates a genetic or chromosomal difficulty. The term *symmetric* refers to the abnormal influence (mostly delayed growth) that affects all organs equally. It is essential to correctly postulate why a symmetric IUGR is often more reversible. Ultrasound-guided intrauterine blood transfusion is often used to improve IUGR in earlier fetal life.

(B) Asymmetric indicates that the head/brain is of the

expected size but the baby's body size and other measures, such as the abdominal circumference, are relatively small. This often indicates poor growth due to inadequate umbilical cord perfusion or anemia.

5.3.4 Skeletal Growth Development

Spinal vertebra development, or lack thereof, can be the first diagnostic clue to spina bifida. Without normal closure, the thecal sac enveloping the spinal cord in the vertebra may be seen to be bulging outside of the vertebrae. Limb development can be noted to have an abnormally small size or an abnormal shape.

5.3.5 Genital Development

Genital development, more than ever before, may be evaluated before delivery. This is made possible by the higher resolution of high-end obstetric ultrasound machines.

5.3.6 Fetal Activity

The fetus's activity can indicate variable states ranging from fetal distress to the infant sleeping. The biophysical profile (BPP) calculates different aspects of fetal motion that indicate fetal well-being. If the fetus has an abnormally low BPP, and especially if the fetus is mature enough to do well outside of the uterus, delivery is often induced or a cesarean is performed.

5.4 Prenatal Ultrasound

Prenatal ultrasound has a continuing rapid pace of development with the advent of 3D and 4D technology. The debate regarding both the usefulness of prenatal ultrasound in terms of the wise expenditure of health care dollars and diagnostic accuracy has lagged behind the explosion of technology. We are charmed by babies, and rightfully so. The development of prenatal ultrasound will no doubt continue and will hopefully benefit humanity.

The subject of prenatal ultrasound allows some discussion about the considerations of technological development. With the development of the prenatal ultrasound technique come widespread expectations and assumptions that may not be true. First, with the demonstrable development of a new technology, it is assumed to work for the intended purpose. It is easy for a layperson to believe that such sophisticated modalities would be 100% accurate and, therefore, worth it. For prenatal ultrasound, it would be predicted that a clear picture would allow for not only accurate prenatal diagnosis but meaningful intrauterine treatment of most conditions. The truth of this assumption is variable, depending, of course, on the condition being evaluated. The greatest accuracy of diagnosis has to do with conditions related to abnormality of amniotic fluid volume. The most discouraging levels of accuracy have to do with the detection of heart defects and isolated congenital abnormalities such as a cleft palate. This subject quickly becomes more complex as one progresses into discussions about issues such as the sensitivity index, cumulative Gaussian distribution, and dimensionless statistics.

5.4.1 Evaluation of Abnormalities

If abnormalities are accurately discovered, the next hurdle is finding an effective treatment. Routinely in sophisticated medical centers, maternal-fetal medicine specialists and perinatologists perform procedures such as amniotic blood transfusions, fetal ureteral valve repair, and removal/replenishment of amniotic fluid volume due to abnormalities. In individual cases, there have been unbelievable advances and actual lives saved with this sophisticated technology. In our ever-cost-sensitive environment, we must balance this incredible cost against the expenditures of limited dollars for patient education.

5.4.2 Evaluation of Embryological Development

The first trimester of pregnancy is when there is the most danger of mishap. Due to the rapid cell division and differentiation of the embryo in unbelievable numbers, even minor "errors" in cellular division may cause the very common event of spontaneous abortion. In the emergency department, it is often an unhappy task for a health professional to explain this issue to patients and their families. It is essential to be careful in doing so due to the tendency of many patients to wonder why a tragedy has occurred and to cast blame on themselves. Without careful explanation, patients who experience spontaneous abortion, or expulsion of an unsuccessful pregnancy through the vagina, may feel that they were at fault when this is not the case.

One can evaluate the earliest pregnancies with many parameters. The serial evaluation of the lab value of the "pregnancy hormone," beta human chorionic gonadotropin (BHCG), over a few days gives an indication of developmental health. Early in pregnancy, the BHCG should double every 2–3 days. If there is a decrease or no increase in the value, this is a most worrisome sign for embryologic death. Often, early fetal demise will not be 100% certain, and the ultrasound and BHCG evaluation are done serially over a few days. Two ultrasounds without a heartbeat at the appropriate fetal age and an inappropriately declining BHCG are often considered diagnostic for intrauterine fetal death and likely spontaneous abortion.

Ultrasound evaluation can most accurately begin in early pregnancy with a longer, higher-frequency transvaginal probe. Returning to the knowledge of ultrasound physics from earlier chapters, it is crucial to realize why a higher-frequency probe is placed through the vagina and near the cervix. The accuracy of the higher-frequency probe allows a marked improvement in resolution when the reflection is only a short distance. At four to five weeks gestation, we first see the fetal yolk sac, fetal pole, gestational sac, and fetal heartbeat. An irregularly shaped gestational sac or lack of a fetal heartbeat when the embryologic or crown-rump length is greater than five weeks are ominous signs of possible embryological death. It is estimated that up to 50% of pregnancies end in spontaneous abortion, often so early in development that the patient does not recognize the pregnancy.

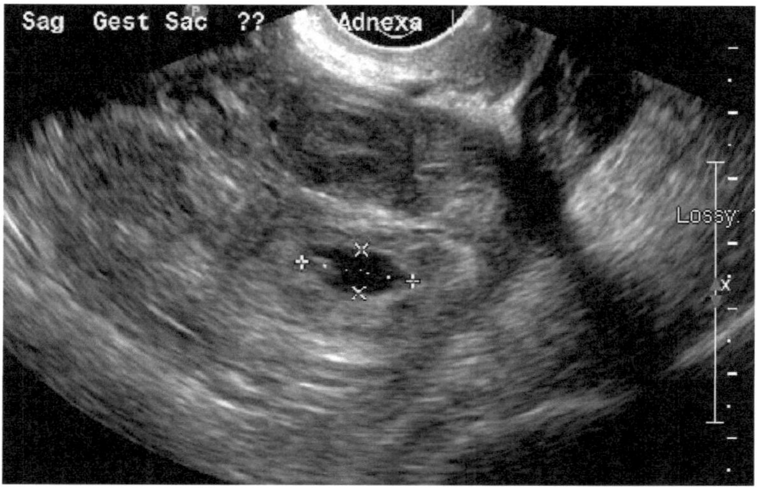

Figure 5-4: An ectopic pregnancy adjacent to the left ovary. Ectopicleftmass by James Heilman, MD licensed under CC BY-SA 3.0

The other common peril in the first trimester of pregnancy is ectopic pregnancy, or pregnancy outside the uterus. As with intrauterine pregnancies, ectopic pregnancies grow and develop at very fast rates. Figure 5-4 shows an ectopic pregnancy adjacent to the left ovary. Most ectopic pregnancies are in the ovaries, fallopian tubes, or uterine horn. In either event, the pregnancies can "outgrow their blood supply," rupturing and leading to maternal hemorrhage and potential death.

Sonographic signs of an intact ectopic pregnancy include enlargement in the area of implantation (even a functioning fetal heartbeat) and an "empty uterus" or lack of an intrauterine pregnancy when the BHCG is greater than 1500 international units. A patient with a ruptured ectopic pregnancy classically presents with sudden-onset unilateral pelvic pain and possibly signs of hemorrhagic shock. On the ultrasound, free fluid may be noted most commonly not only in the pouch of Douglas but also in the paracolic gutters or Morrison's pouch. This is often a surgical emergency; the bleeding must be stopped before maternal death occurs from hemorrhage.

Gestational dating refers to the estimation of the gestation of the pregnancy. This is somewhat helpful when the medical provider evaluates issues such as whether to try to delay a patient in preterm labor or if induction of a patient who continues pregnancy more than 10 days after the estimated date of confinement (EDC) is indicated. Both of these issues have several determining factors. For preterm labor, infant mortality significantly decreases as delivery is delayed in many cases. For postterm situations, having a healthy term fetus more than 10 days after the EDC may pose unnecessary risks. These are complex decisions based on such data. To further complicate the issue, gestational dating may have variable accuracies. Some individuals have little or no prenatal care (i.e., some show up only to the emergency department when they are about to deliver or deliver at home without ever being evaluated by a medical provider). If one does have prenatal care and prenatal ultrasound, gestational dating accuracy remains variable. The facet that introduces the most variability is the gestational age at which the scan is performed. It has been traditionally noted that gestational dating accuracy has an 8% variability of the maturity of the pregnancy.[5] An example is that a 45-day gestation would have an accuracy of ±3.5 days, and a pregnancy of approximately 250 days would have an accuracy of ±20 days. This variability may decrease as ultrasound machines become more sophisticated. The principle, however, continues to make sense that as pregnancies mature, the genetic and environmental issues that make us all different have a greater influence on growth.

5.5 The Second and Third Trimester

At approximately 14 weeks gestation, evaluation of organogenesis becomes more useful. As there is fetal growth, there is predictably more accuracy but fewer chances for potential interventions that may help as pathologic damage becomes irreversible. Organogenesis evaluation may differ according to the medical circumstances. Diagnostic capabilities vary according to the operator and machine.

The head, heart chambers, lungs, abdominal organs, genitals, and bones may be more successfully evaluated in the second trimester. While diagnostic accuracy improves, individual organs remain the most difficult to evaluate. Detecting some prenatal heart defects is an area of great interest and some improvement. The most common advantage of prenatal diagnosis of congenital heart defects is that if the condition is known, in more

5. Committee Opinion No 700: Methods for Estimating the Due Date. Obstet Gynecol. 2017 May;129(5):e150-e154. doi: 10.1097/AOG.0000000000002046. PMID: 28426621.

developed countries, the delivery can often be done at a tertiary medical center where there are maternal-fetal medicine specialists, pediatric cardiologists, pediatric surgeons, neonatal ventilators, and other sophisticated options for immediate treatment.

Warda et al.[6] are originally credited with developing regression analysis values to use fetal growth parameters to estimate gestational age. Due to an interesting phenomenon known as "organ sparing," some organs are more reflective of accurate gestational dating than others. If there is placental insufficiency, the head will continue to grow more normally, but abdominal growth will lag. This increasing discrepancy between fetal head and abdominal growth may lead to the diagnosis of IUGR and indicate induced delivery of a small or preterm infant.

5.5.1 Head Evaluation

The head measurement is first determined by the biparietal diameter and occipital-frontal circumference. There is a flow pattern of cerebrospinal fluid (CSF) through the brain's ventricular system and into the spinal cord through the foramen of Magendie and the foramen of Luschka. An obstruction of this flow causes increased back pressure, dilation of the ventricles, and pathologic pressure of the surrounding brain tissue. Many other conditions other than stenosis of the foramina cause hydrocephalus. This abnormality may be seen on prenatal ultrasound first as an increased head size and then as an increased ventricular size. Figure 5-5 shows an ultrasound image of fetal hydrocephalus.

6. Warda AH, Deter RL, Rossavik IK, Carpenter RJ, Hadlock FP. Fetal femur length: A critical reevaluation of the relationship to menstrual age. Obstet Gynecol. 1985 Jul;66(1):69–75. PMID: 3892390.

Hydrocephalus intervention is most often treated with postdelivery shunt placement, which involves placing a tube in the cerebral ventricles to bypass the obstruction and drain the excess fluid into other body cavities, such as the abdomen, where the fluid may be reabsorbed.

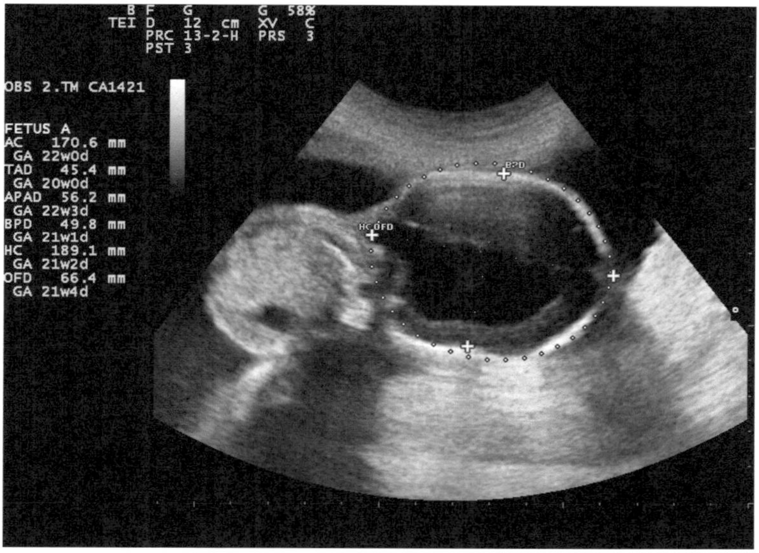

Figure 5-5 Ultrasound image of fetal hydrocephalus. Aorta duplication artifact 131206105958250c by Nevit Dilmen licensed under CC BY-SA 3.0

Other than abnormalities in CSF flow, there may be other abnormalities that are represented by an abnormal prenatal head ultrasound. These abnormalities have a wide range of etiologies, including genetic abnormalities, congenital abnormalities, and abnormal growth in an otherwise normal fetus. Genetic abnormalities have to do with abnormal chromosomes primarily due to mutation. Congenital abnormalities usually begin in the prenatal period and are evident at birth.

Cardiac ultrasound accurately predicts congenital abnormalities for many more subtle conditions. The most common congenital heart disease is patent ductus arteriosus, where the fetal ductus arteriosus that diverts the circulation away from the prenatal lungs and into the aorta does not close upon birth. This, of course, is never diagnosed prenatally. Rare but hazardous and sometimes fatal conditions such as transposition of the great vessels or defects of the central cardiac wall (called the septal wall) may be undetected by screening ultrasound exams. The complexity increases when there are combinations of heart defects such as tetralogy of Fallot, including pulmonary stenosis, right ventricular hypertrophy, ventricular septal defect, and an aorta that receives blood from both the right and left ventricle. In developed countries, when a fetal cardiac ultrasound is suspected from a routine prenatal ultrasound, a targeted ultrasound using more specialized machines and more specialized operators is subsequently performed. There may also be noncardiac prenatal ultrasound signs of heart disease, such as fetal hydrops (swelling) for fetal congestive heart failure.

5.5.2 Abdominal Exam

Most intrauterine conditions that are treatable prior to delivery are identified in the abdominal exam. The umbilical cord is utilized as a site of intrauterine blood transfusions (by maternal-fetal specialists) to correct fetal anemia and hypovolemia. This procedure is done under real-time ultrasound and may be genuinely lifesaving as well as preventative of anomalies such as Potter's sequence.

The fascinating development of the normal kidneys involves the division and migration of one central organ into a right and left kidney. Abnormalities in this separation and division include the persistence of the larger central organ, called a "horseshoe kidney." Difficulties in renal function, such as nonfunction, may be suspected when the kidney appears small or atrophic. If a kidney is obstructed but functioning, "backward pressure" from the fetal urine may cause the kidney to enlarge, a condition known as hydronephrosis.

Some urinary tract obstructions may be corrected by intrauterine surgery. Posterior urethral valves in the male are the most common correction. When successful, relieving the obstruction prevents further renal damage and restores the average amniotic fluid volume to prevent complications from abnormal pressure.

5.5.3 Amniotic Fluid

Evidence of the normal flow of amniotic fluid is indicated by amniotic fluid volume. Contributions to the average amniotic fluid volume are mostly from fetal urine production, with some contributions from umbilical cord filtrate. Detraction from the amniotic fluid volume is mainly from fetal swallowing and fetal aspiration of the amniotic fluid into the lungs.

Oligohydramnios, a condition caused by reduced amniotic fluid volume from the reduced contribution of urine, creates well-known fetal anomalies called Potter's sequence. This condition is described as including particular fetal facial characteristics, fetal limb abnormalities, and pulmonary hypoplasia. The limb abnormalities are from abnormal pressure.

Polyhydramnios may be from inadequate fetal swallowing of amniotic fluid, inadequate fetal aspiration of amniotic fluid, or fetal gastrointestinal obstruction that prevents amniotic fluid from progressing once swallowed.

5.5.4 Femur Bone Length

The femur length is often considered to be the most reliable indicator of gestational age. Congenital abnormalities of the femur that make this statement not 100% true are rare and often apparent, as the femur abnormality coexists with other skeletal difficulties.

5.6 Tertiary Ultrasound Exams

In first discussing the more common screening prenatal ultrasound exams, we have made mention of more targeted exams. Thoracic measurements, measurement of bones other than the femur, more targeted cardiac exams, more central nervous system ventricular evaluations, and fetal neck skin thickness measurements are made. For example, Figure 5-6 shows ultrasound images of a fetus's face and arms during week 20.

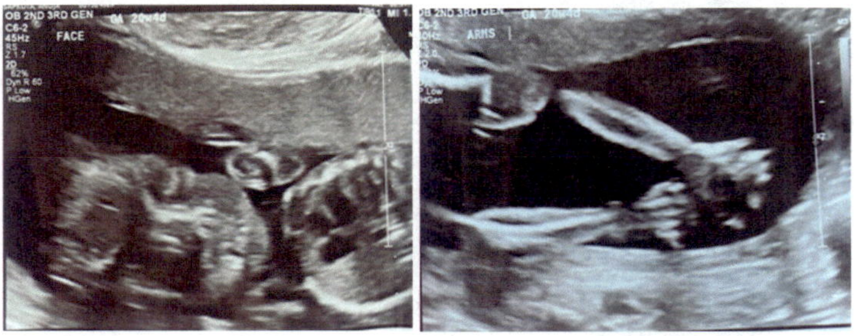

Figure 5-6: Ultrasound images of a fetus's face and arms during week 20.

More specific investigation of suspected abnormalities may indicate some conditions that can be corrected prenatally by thrilling advanced treatments such as umbilical cord blood transfusions, correction of amniotic fluid abnormalities, relief of posterior ureteral valve abnormality, or other intrauterine surgeries. As discussed earlier, almost all of these procedures are initiated by the targeted ultrasound.

In 2023, as in every other year in the history of humanity, it remains true that the results of the most sophisticated events in fetal development can be influenced by some of the simplest yet sometimes neglected measures: receiving good prenatal care, ceasing maternal smoking or other substance use, and taking prenatal vitamins. We believe that it is important for future clinicians reading this to never tire of reminding patients to optimize their developing infants' likelihood of being healthy.

5.7 *Self-Assessment*

1. Define *sensitivity* and *specificity* in the context of prenatal ultrasound.
2. In early pregnancy, how often should the beta human chorionic gonadotropin (BHCG) double, typically?
3. What are three forms of abnormalities that can be identified by performing a prenatal head ultrasound?
4. What measurements does the tertiary ultrasound exam include?

5.8 *Further Readings*

1. Acton CM. The use of ultrasound in pregnancy. Aust Fam Physician. 1997 Mar;26(3):239–43. PMID:

9078656.

2. Whitworth M, Bricker L, Mullan C. Ultrasound for fetal assessment in early pregnancy. Cochrane Database Syst Rev. 2015 Jul 14;2015(7):CD007058. doi: 10.1002/14651858.CD007058.pub3. PMID: 26171896; PMCID: PMC6464767.
3. AIUM-ACR-ACOG-SMFM-SRU Practice Parameter for the Performance of Standard Diagnostic Obstetric Ultrasound Examinations. J Ultrasound Med. 2018 Nov;37(11):E13–E24. doi: 10.1002/jum.14831. Epub 2018 Oct 11. PMID: 30308091.
4. Salomon LJ, Alfirevic Z, Bilardo CM, Chalouhi GE, Ghi T, Kagan KO, Lau TK, Papageorghiou AT, Raine-Fenning NJ, Stirnemann J, Suresh S, Tabor A, Timor-Tritsch IE, Toi A, Yeo G. ISUOG practice guidelines: Performance of first-trimester fetal ultrasound scan. Ultrasound Obstet Gynecol. 2013 Jan;41(1):102–13. doi: 10.1002/uog.12342. Erratum in: Ultrasound Obstet Gynecol. 2013 Feb;41(2):240. PMID: 23280739.

6. Musculoskeletal Sonography

6.1 Learning Objectives

After reviewing this chapter, you should be able to do the following:

1. Understand how to perform diagnostic exams for various musculoskeletal (MSK) anatomical structures.
2. Identify and attempt to correct related artifacts in MSK examinations.
3. Begin to identify some abnormal MSK structures.

6.2 Introduction

Ultrasound can be used to evaluate different anatomic MSK structures for diagnostic and therapeutic purposes. In particular, protocols have been developed to evaluate different joint structures of the upper and lower extremities. A complete MSK ultrasound of an extremity incorporates real-time scanning of a specific joint, including muscles, tendons, ligaments, or other structures and any abnormalities. A limited MSK ultrasound is a focused evaluation of a specific anatomic structure, such as a tendon or muscle injury. To develop competence in MSK evaluation, being familiar with anatomy, function, and pathology is imperative. This chapter will introduce fundamental structures we find in routine MSK sonography.[1]

6.3 Performing a Diagnostic Exam

In general, transverse and longitudinal planes (also called sagittal planes) should be obtained of all key anatomic structures and pathologies when performing a diagnostic evaluation. It is always important to know your orientation when visualizing the images. Transducers usually have an indicator on one side at the end of the probe, such as a light, knob, or notch corresponding to the right side of the screen's image. If you are unsure, just touch one edge of the probe with your finger, and look at the screen to see if you are on the right or left side before imaging. During the evaluation, compare static images, dynamic images, Doppler evaluation, and possible contralateral evaluation. Also, document any masses or fluid collections, such as bursal distension, by indicating the location, size, shape, echotexture, compressibility, and presence or absence of flow with Doppler.[2]

Image optimization is obtained by selecting the proper transducer with appropriate frequencies. In general, a high-frequency linear transducer (10 MHz or higher) will be the appropriate selection for the evaluation of most joints. However, sometimes a curvilinear transducer gives better imaging in larger joints and in the evaluation of most adult hips, which usually require

1. Jacobson JA. Fundamentals of Musculoskeletal Ultrasound. 3rd ed. [place unknown]: Elsevier Saunders; 2017. 472 p.
2. Jacobson JA. Fundamentals of Musculoskeletal Ultrasound. 3rd ed. [place unknown]: Elsevier Saunders; 2017. 472 p.

deeper penetration for visualization. Another exception would be the evaluation of superficial detailed structures such as the pulley system of the digits in the hand. A hockey stick transducer (>10 MHz) would be more appropriate for better resolution. Figure 6-1 shows some commonly used transducers in musculoskeletal ultrasound assessment. When evaluating a joint, it is helpful to start by directing your angle of insonation to the bony cortex, which is usually the most distal and hyperechoic structure, to avoid anisotropy.[3]

Linear transducer	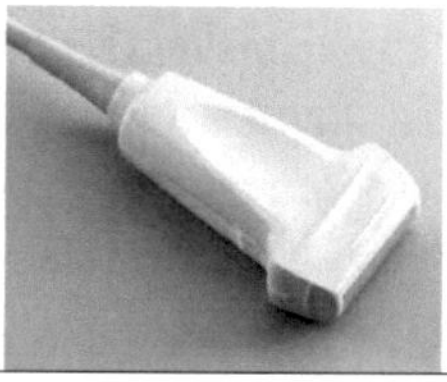
Curvilinear transducer	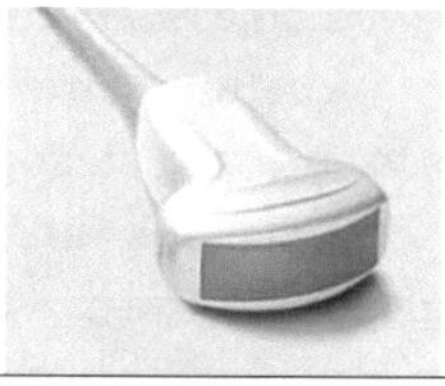
Hockey stick transducer	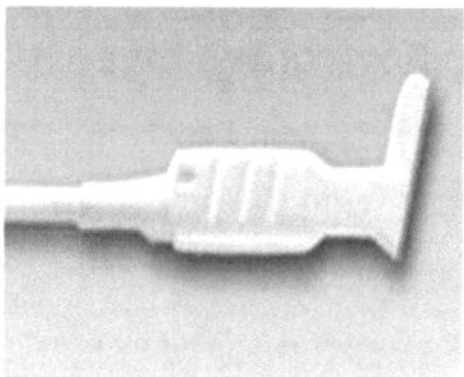

Figure 6-1: Some examples of commonly used transducers in musculoskeletal ultrasound assessment.

3. Jacobson JA. Fundamentals of Musculoskeletal Ultrasound. 3rd ed. [place unknown]: Elsevier Saunders; 2017. 472 p.

6.4 Identifying Abnormal Structures

The normal sonographic appearance of MSK structures often has characteristic ultrasound images that are best visualized in the longitudinal plane. For example, tendons usually appear as a hyperechoic fibrillar echotexture. Ligaments are similar but more compact and connect two osseous structures. Muscle tissue appears more hypoechoic with septations, a pennate (featherlike) appearance in the longitudinal plane, and a starry-night appearance in the transverse plane with dynamic maneuvers. Bone is usually very hyperechoic. It creates a significant acoustic impedance mismatch and therefore is very reflective and appears bright white (hyperechoic) on the image. Adipose tissue and cartilage tend to be hypoechoic. Nerves tend to have both a hypoechoic and hyperechoic honeycomb appearance. The location and function of the structures are always helpful when determining normal anatomy and pathology.

Injuries, inflammation, or infections are divided into *acute* and *chronic* and can affect any musculoskeletal structure. Acutely, the sonographic structures tend to be hyperechoic, with possible hypertrophy, hypervascularity, fluid, and disruption of fibers in structures such as tendons. Chronic sonographic images tend to be more hypoechoic, with possible atrophy, scarring, and areas of calcification. Bone abnormalities can also be seen in acute and chronic processes. The normal bone cortex is smooth, uniform, and hyperechoic. A bone fracture can be visualized as a discontinuity of the bone cortex and disruption of the cartilage. Arthritis can have characteristic bone images such as crystal deposits on the cartilage surface in gout and synovial hypertrophy with bone erosions in rheumatoid arthritis.

Sonographic artifacts are not uncommon with MSK ultrasound, some of which have been discussed in earlier chapters. It is vital that the sound beam is perpendicular to the anatomic structure being visualized, or anisotropy can be encountered and give false information. To correct for anisotropy, performing a heel-to-toe maneuver on the long axis and toggling the transducer on the short axis are often helpful to finely tune the ultrasound image.

6.5 Shoulder Sonography

Figure 6-2 shows the anterior view of the shoulder anatomy. The shoulder is one of the most accessible joints to perform a comprehensive ultrasound evaluation. An ultrasound evaluation can be as reliable as an MRI for a rotator cuff tear. A complete shoulder evaluation should include the rotator cuff's tendons and muscles, including the subscapularis, supraspinatus, infraspinatus, and teres minor. Also, examine the biceps brachii (with dynamic maneuvers, if indicated for subluxation, dislocation, or impingement), the acromioclavicular joint, the suprascapular nerve (in the suprascapular notch and the spinoglenoid notch), and the posterior glenohumeral joint.[4] Evaluating each anatomic structure in the transverse and longitudinal planes is essential.

4. Moore RE. Protocol for the shoulder, elbow, wrist, hand, hip, knee, ankle, and foot [DVD]. St Petersburg (FL): Gulfcoast Ultrasound Institute; 2010, 2013.

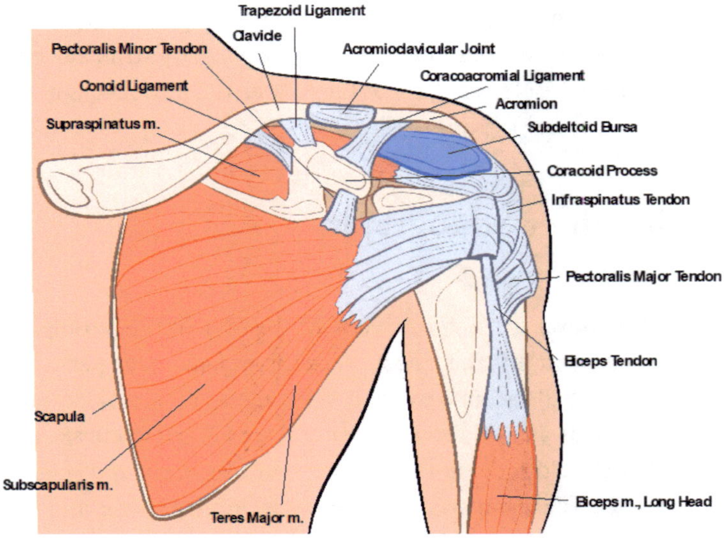

Figure 6-2: Anterior view of shoulder anatomy.

For examination of the patient's shoulder, developing an approach that allows for the best visualization with dynamic maneuvers is helpful. One approach would be to start by standing in front of the seated patient with their arm at their side, the elbow at 90 degrees flexion, the forearm in supination, and the ultrasound machine on one side of the patient for exam visualization.

The long-head biceps tendon is the first structure to be evaluated, and it works as a good reference point for the anterior shoulder evaluation. The origin of the long-head biceps is the supraglenoid tubercle of the scapula, and the insertion is the radial tuberosity and bicipital aponeurosis. It is innervated by the musculocutaneous nerve. Its action is flexion and supination of the forearm at the elbow joint and flexion of the arm at the shoulder joint. First, look at the transverse position within the bicipital groove of the humeral head with a linear transducer.

The biceps tendon should have a bright, dense, ovoid, and

bristle-like appearance. It should be assessed from proximal to distal in the transverse and longitudinal planes, as shown in Figure 6-3. It is essential to evaluate the most proximal area where the biceps tendon courses over the humeral head because this is a common site for pathology. Also, fluid distension within the bicipital tendon sheath often indicates shoulder pathology, since part of it communicates with the shoulder joint. Continue the evaluation distally until the fibrous-appearing band of the pectoralis major inserted into the proximal humerus is visualized. This is sometimes where a bicipital tendon tear can be found separated from the muscle after an injury. Dynamic maneuvers, both active and passive, can be very helpful in the evaluation.

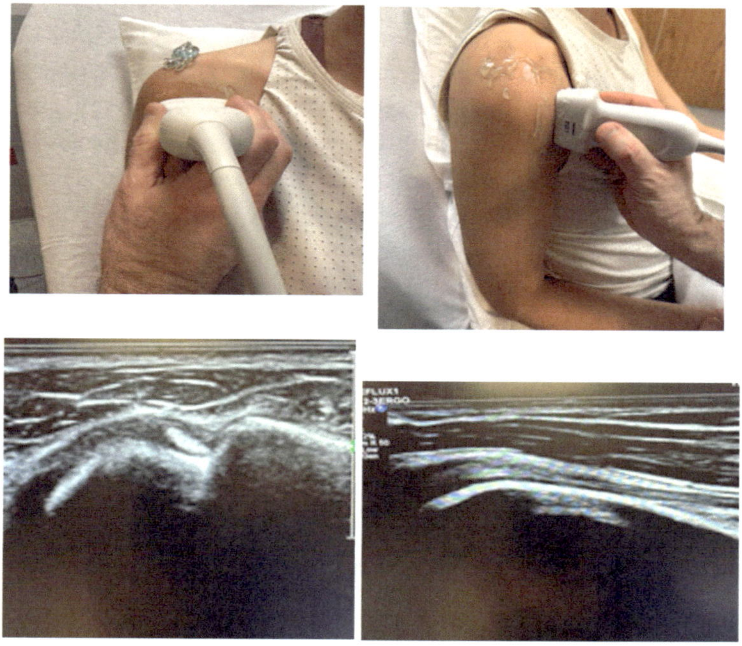

Figure 6-3: Structure of the biceps tendon viewed in the transverse and longitudinal planes.

After looking at the bicipital tendon/muscle, return to the point of reference within the bicipital groove of the humerus in the transverse plane. Next, evaluate the subscapularis, which originates at the subscapular fossa and inserts into the lesser tubercle of the humerus. It is innervated by the upper superior and lower inferior subscapular nerves, and the action is for internal rotation of the humeral head. It prevents anterior displacement of the humerus. First, start by moving the transducer medially to the lesser tuberosity to evaluate the rotator interval, which is the space between the anterior margin of the supraspinatus tendon and the superior margin of the subscapularis tendon. The subscapularis tendon/muscle is evaluated by a passive range of motion with external rotation, as shown in Figure 6-4. This brings the subscapularis into the longitudinal (sagittal) plane as it rotates over the humerus. During this dynamic maneuver, also evaluate for coracoid impingement. A limited view of the anterior glenohumeral joint can also be evaluated in this position. The probe is then rotated 90 degrees clockwise in the transverse plane. In this view, the subscapularis will have a characteristic vertical hypoechoic segmented appearance secondary to the musculoskeletal junction, which is usually normal anatomy and not a tear. The evaluation should again include any evidence of effusion, synovial hypertrophy, or tearing.[5],[6]

5. Jacobson JA. Fundamentals of Musculoskeletal Ultrasound. 3rd ed. [place unknown]: Elsevier Saunders; 2017. 472 p.
6. Msksono.org: Musculoskeletal Ultrasonography [internet]. [place unknown]; c2022 [cited 2023 Oct 28]. Available from: https://msksono.org/

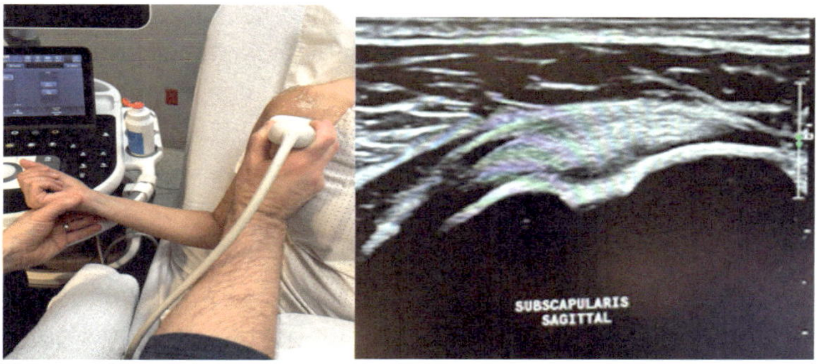

Figure 6-4: Examination of the subscapularis tendon in the sagittal plane.

The next anatomic structure to evaluate in the anterior position of the patient is the acromioclavicular (AC) joint, shown in Figure 6-5. The most straightforward approach is to palpate the AC joint and place the linear transducer on top of it in the transverse plane. Evaluate for widening, such as in a tear or effusion, which is sometimes indicative of rotator cuff pathology.

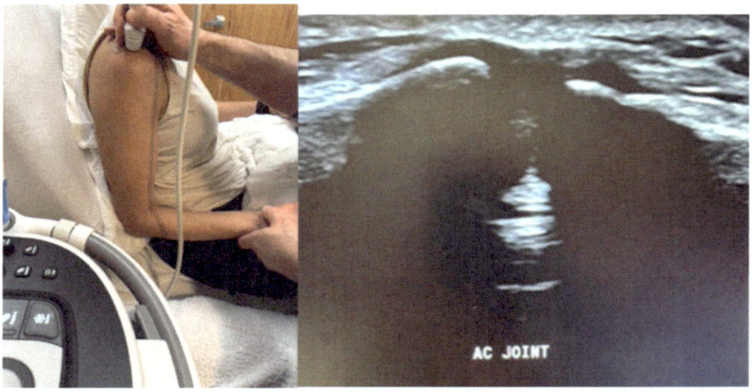

Figure 6-5: Examination of the acromioclavicular joint in the sagittal plane.

The next rotator cuff to be evaluated is the supraspinatus, which originates in the supraspinatus fossa and is inserted on the superior facet of the greater tubercle of the humerus. Innervation is by the suprascapular nerve; its action is the abduction of the arm and stabilization of the glenohumeral joint. The best position for the patient to be in is called the *modified crass position*. In this position (which involves extension, adduction, and internal rotation), the patient is sitting upright with the palm of their hand on the ipsilateral hip and the elbow flexed and pointed posteriorly. This brings the supraspinatus out from under the cover of the acromion. Over 90% of rotator cuff injuries involve the supraspinatus.[7], [8] First, evaluate the supraspinatus tendon in the longitudinal plane, as shown in Figure 6-6. This image is the most essential view and should have a bird's beak appearance. Next, include the transverse plane view. Evaluate the bony cortex, hyaline cartilage, supraspinatus tendon/muscle, peribursal fat, and the subacromial bursa. Pooling of fluid within the subacromial bursa or restrictive motion of the supraspinatus tendon could indicate subacromial impingement.[9]

7. Jacobson JA. Fundamentals of Musculoskeletal Ultrasound. 3rd ed. [place unknown]: Elsevier Saunders; 2017. 472 p.
8. Msksono.org: Musculoskeletal Ultrasonography [internet]. [place unknown]; c2022 [cited 2023 Oct 28]. Available from: https://msksono.org/
9. Jacobson JA. Fundamentals of Musculoskeletal Ultrasound. 3rd ed. [place unknown]: Elsevier Saunders; 2017. 472 p.

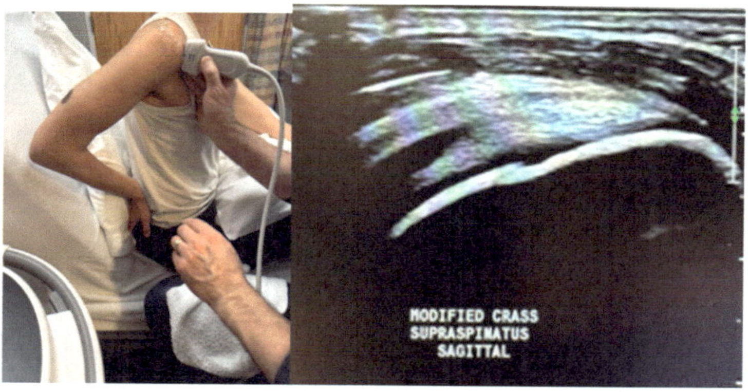

Figure 6-6: Supraspinatus in the longitudinal (sagittal) view showing the bird's beak appearance.

It is vital to evaluate the tears of the supraspinatus with the correct description. First, determine if it is a full-thickness tear extending from the articular to the bursal surface or a partial-thickness tear. Partial-thickness tears involve the articular or bursal surface or are localized within the tendon, not extending to either surface. This is called an *intrasubstance tear*. When evaluating the diameter of the tear, measure along the long and short axis.

Posterior cuff imaging is evaluated next by facing the posterior shoulder, palpating the scapular spine, and placing the transducer below it in an oblique axial plane angled superiorly toward the humeral head. A curvilinear probe is sometimes needed for better penetration, since the posterior shoulder is a deeper structure. The infraspinatus and teres minor tendons are first evaluated in the longitudinal plane from the scapular fossa's origin to the humerus's greater tuberosity, as shown in Figure 6-7. The origin of the infraspinatus is the infraspinatus fossa of the scapula, and the insertion is on the middle facet of the greater tuberosity of the humerus. The suprascapular nerve innervates it, and its action is for external rotation and abduction of the arm at the shoulder joint with stabilization of the shoulder joint. The teres minor originates on the lateral border of the scapula and inserts onto the inferior facet of the greater tuberosity of the humerus. It is innervated by the axillary nerve and functions similarly to the infraspinatus.

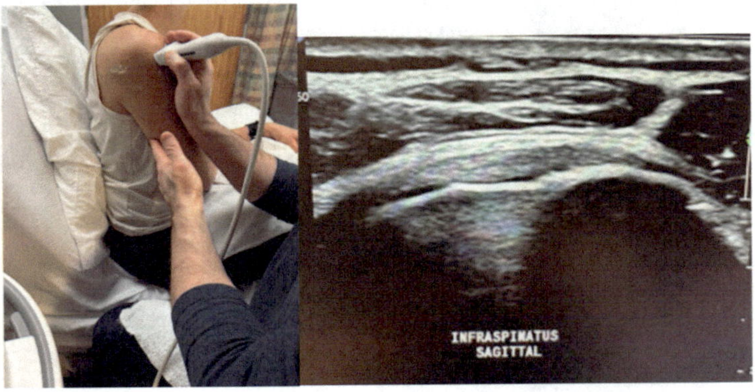

Figure 6-7: The infraspinatus in the longitudinal (sagittal) plane.

Next, evaluate the suprascapular nerve in the suprascapular notch and the spinoglenoid notch. Sometimes, turning on the Doppler to better visualize the suprascapular artery is helpful, and right next to it is the suprascapular nerve.

Finally, evaluate the posterior glenohumeral joint, as shown in Figure 6-8. Look for joint effusion, cortical irregularities, and osteophytes, and evaluate the posterior labrum for cysts or tears. Also, this is a good approach for intra-articular glenohumeral joint injections using a posterior approach. This completes the shoulder evaluation.

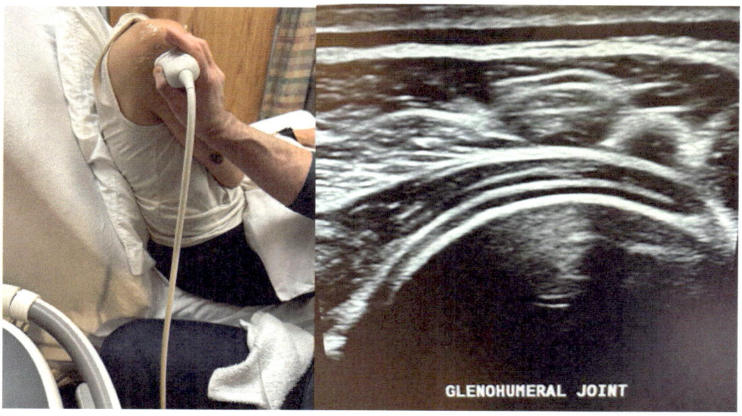

Figure 6-8: The posterior glenohumeral joint on ultrasound.

6.6 Elbow Sonography

The entire elbow examination is usually accomplished with a linear transducer. Like all the other joints, the elbow is best evaluated in a quadrant approach: anterior, medial, lateral, and posterior. Figures 6-9 and 6-10 show the anatomical structures of the elbow.

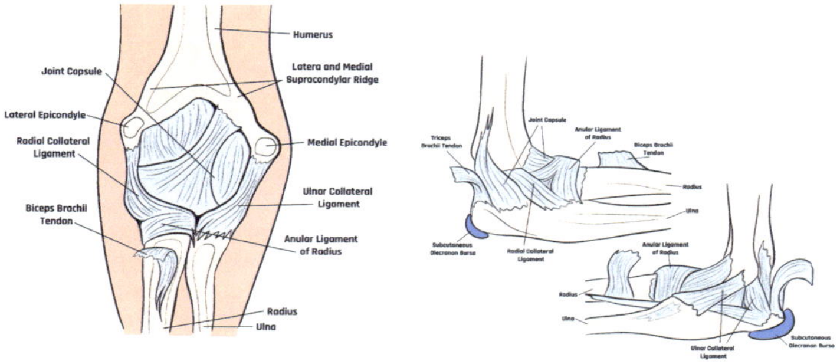

Figure 6-9: Anatomic bone and ligamentous structures of the anterior elbow.

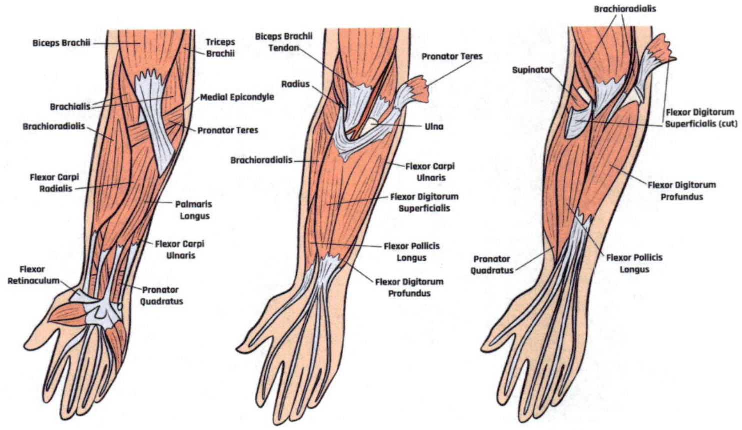

Figure 6-10: Anatomic muscle and tendon structures of the anterior elbow.

Anteriorly, look at the joint space for narrowing, cortical and cartilage irregularities, synovial hypertrophy, and effusion. Also evaluate the brachialis, biceps, median and radial nerves, and the brachial artery, as shown in Figure 6-11.

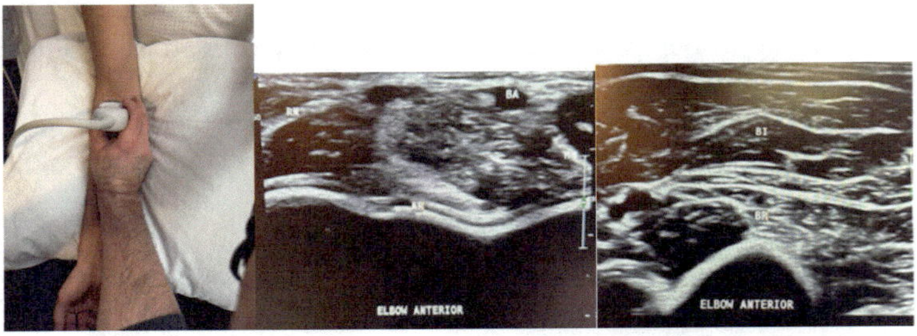

Figure 6-11: Side-by-side pictures of the transducer position on the elbow and anterior transverse elbow imaging of the anterior recess (AR), radial nerve (RN), brachial artery (BA), median nerve (MN), biceps (BI), and brachialis (BR) just proximal to the elbow crease. The abbreviations given here and labeled on the ultrasound images represent the corresponding structures.

Next, the medial elbow evaluation is performed with the elbow in partial extension and the probe in a longitudinal axis, as shown in Figure 6-12. Evaluate the anterior band of the ulnar collateral ligament (UCL). It will have a characteristic triangular homogenous appearance as it spans from its attachment proximally to the humeral trochlea and distally to the olecranon. The common flexor tendon is superficial to the UCL and evaluated carefully at the insertion point of the medial epicondyle, since this is the site of medial epicondylitis. The pronator teres should also be examined for any evidence of tears, effusion, or synovial hypertrophy.[10,11]

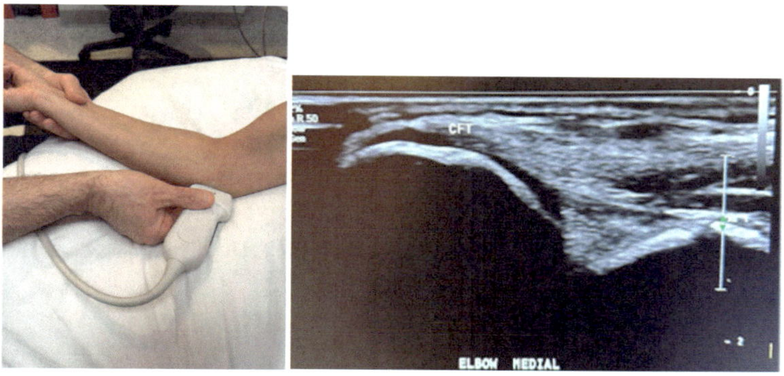

Figure 6-12: Side-by-side pictures of the transducer position and a longitudinal image of the medial elbow showing the common flexor tendon (CFT).

The lateral elbow is next approached with the elbow flexed at 90 degrees and the ipsilateral hand resting in pronation, as shown

10. Jacobson JA. Fundamentals of Musculoskeletal Ultrasound. 3rd ed. [place unknown]: Elsevier Saunders; 2017. 472 p.
11. Msksono.org: Musculoskeletal Ultrasonography [internet]. [place unknown]; c2022 [cited 2023 Oct 28]. Available from: https://msksono.org/

in Figure 6-13. A longitudinal probe placement is performed to evaluate the bony margins of the capitulum of the humerus and the radial head. Evaluate the radial collateral ligament complex and the common extensor tendon (CET). Closely evaluate the attachment of the CET to the lateral epicondyle, the site of lateral epicondylitis.

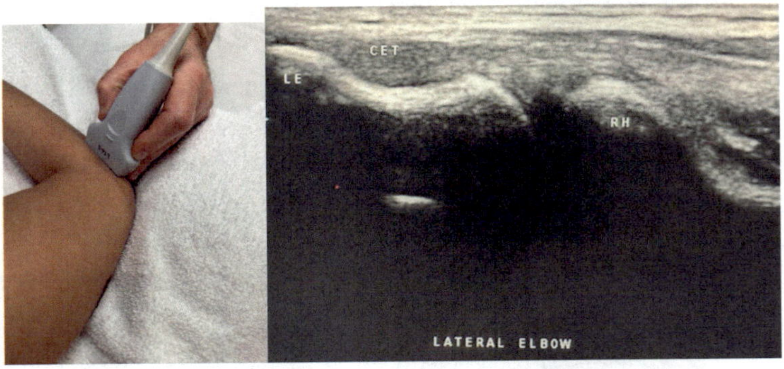

Figure 6-13: Side-by-side pictures of the transducer position and a longitudinal image of the lateral elbow showing the common extensor tendon (CET), lateral epicondyle (LE), and radial head (RH). The abbreviations given here and labeled on the ultrasound image represent the corresponding structures.

Finally, the posterior evaluation is performed with the elbow at approximately 90 degrees of flexion, as shown in Figure 6-14. Evaluate the triceps muscle and tendon, olecranon bursa, and the ulnar nerve within the groove between the medial epicondyle and the olecranon of the ulna. This can be a site for entrapment of the ulnar nerve, which should have an area of 7 mm or less.[12],[13]

12. Msksono.org: Musculoskeletal Ultrasonography [internet]. [place unknown]; c2022 [cited 2023 Oct 28]. Available from: https://msksono.org/
13. Jacobson J, Kissin E, Lento P, Mazzola T, Moore RE, Shapiro S.

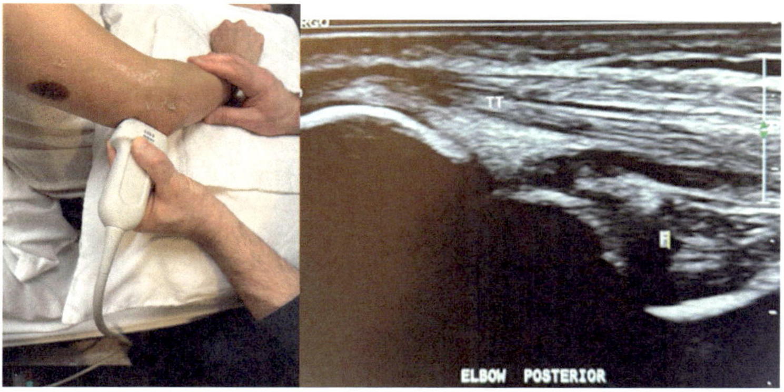

Figure 6-14: Side-by-side pictures of the transducer position and a posterior longitudinal image of the elbow showing the triceps tendon (TT) and the olecranon fossa (F). The abbreviations given here and labeled on the ultrasound image represent the corresponding structures.

6.7 Wrist and Hand Sonography

The wrist and hand anatomy involve superficial structures; therefore, the best approach is to use a high-frequency hockey stick transducer for better resolution. Figure 6-15 shows the anatomical structure of the wrist and hand.

Introduction to Musculoskeletal Ultrasound [DVD]. St Petersburg (FL): Gulfcoast Ultrasound Institute; 2014 Jan.

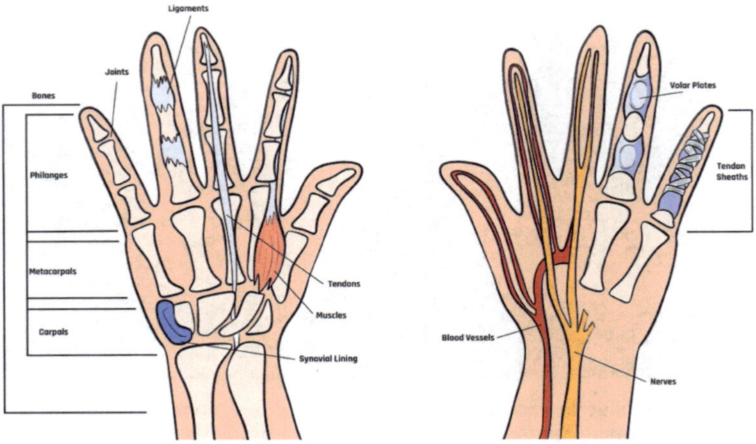

Figure 6-15: Schematic of the wrist and hand anatomy.

Start with the palm of the hand facing down. Identify Lister's tubercle (see Figure 6-17) on the dorsum of the distal radius by digital palpation, and place the transducer on top of it in a transverse plane, as shown in Figure 6-16. This bony prominence separates the second and third extensor tendon compartments of the six in the wrist and helps with identification and orientation. Just radial to Lister's tubercle is the second compartment containing the extensor carpi radialis brevis and the extensor carpi radialis longus.

With further radial movement, the first compartment on the side of the wrist is identified, which contains extensor pollicis brevis and abductor pollicis longus tendons. The first compartment is the site of de Quervain's tenosynovitis. Evaluate each compartment from proximal to distal in both planes with active and passive dynamic maneuvers depending on the clinical concerns. Superficial to the compartments is the extensor retinaculum.

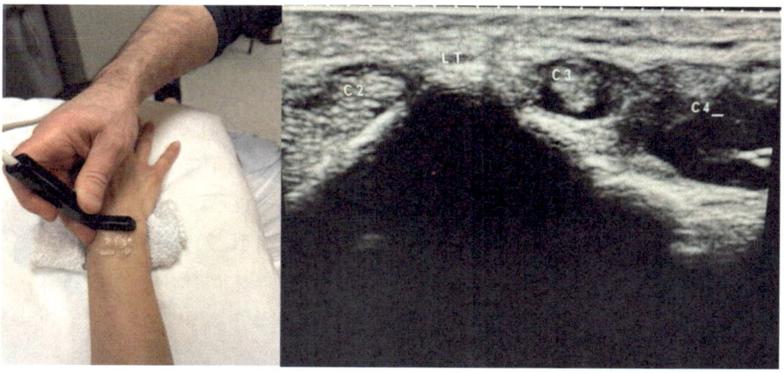

Figure 6-16: Side-by-side pictures of the transducer position and dorsal transverse wrist showing the second extensor tendon compartment (C2), Lister's tubercle (LT), third extensor tendon compartment (C3), and fourth extensor tendon compartment (C4ETC). The abbreviations given here and labeled on the ultrasound image represent the corresponding structures.

The third compartment is on the ulnar aspect of Lister's tubercle and contains the extensor pollicis longus, as shown in Figure 6-17. Moving further toward the ulna, next to the third compartment, is the fourth compartment, which contains multiple extensor digitorum tendons and extensor indicis. Compartment five contains the extensor digiti minimi, and the sixth compartment contains the extensor carpi ulnaris, as shown in Figure 6-17. Evaluate each for any pathology.

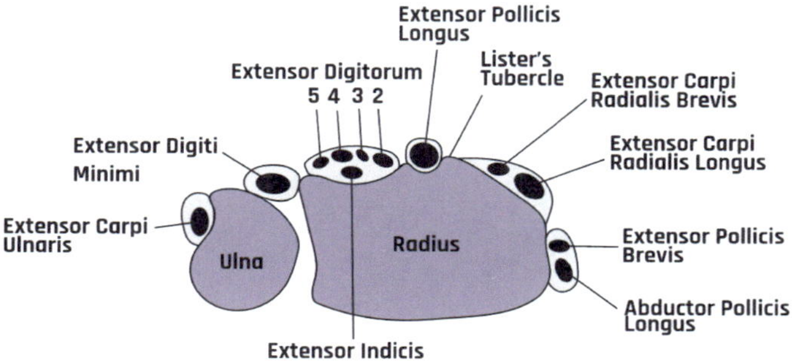

Figure 6-17: Schematic of the cross-sectional view of the wrist.

After evaluating each compartment, return to Lister's tubercle in the transverse plane, and move the probe distally to the radiocarpal joint. The bone just distal to the radius is the scaphoid, and the lunate bone is next to the scaphoid in the ulnar direction. Between the dorsal aspects of both bones is a triangular area that represents the scapholunate ligament, which has a compact hyperechoic fibrillar echotexture, as shown in Figure 6-18. This is a common site for injuries from falls involving extended wrists that could result in a tear of the scapholunate ligament. It is also a common site for ganglion cysts.

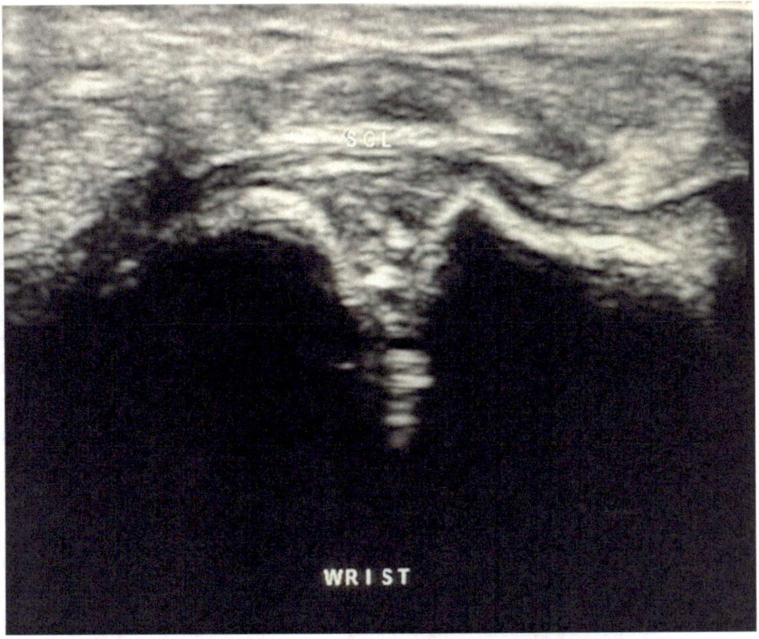

Figure 6-18: Dorsal wrist image of the scapholunate ligament (SCL).

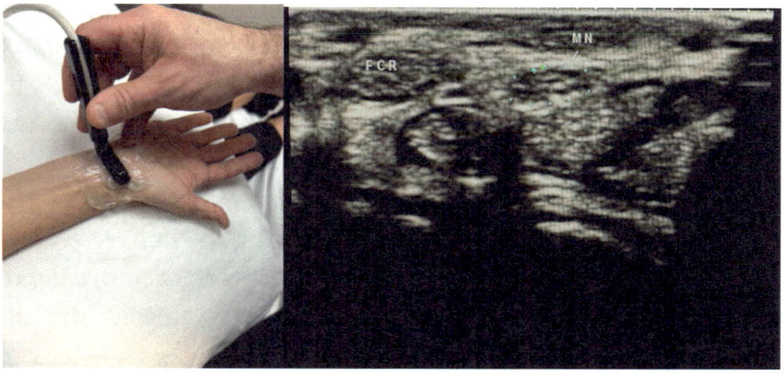

Figure 6-19: Side-by-side pictures of the transducer position and volar wrist showing the median nerve (MN) and flexor carpi radialis (FCR). The abbreviations given here and labeled on the ultrasound image represent the corresponding structures.

Now rotate the hand to evaluate the volar aspect, as shown in Figure 6-19. Evaluate the median nerve, flexor tendons, volar joint recesses, flexor carpi radialis, palmaris longus, and radial artery and the flexor tendons, pulleys, volar plates, collateral ligaments, and joint recesses of the fingers as clinically indicated. The median nerve is found between the flexor carpi radialis and palmaris longus. Place the transducer between these two tendons in the distal wrist crease in the transverse plane, and move the probe proximally as the honeycomb appearance of the median nerve courses radial to the flexor tendons and then moves ulnar and deep between the flexor digitorum superficialis and profundus. If the median nerve has a cross-sectional area of 12 mm^2 or greater, this suggests carpal tunnel syndrome. Also, a 2 mm^2 or greater difference in the cross-sectional area of the median nerve measured proximally at the level of the pronator quadratus and distally at the level of carpal tunnel has a 99% accuracy for carpal tunnel syndrome.[14]

14. Kamolz LP, Schrögendorfer KF, Rab M, Girsch W, Gruber H, Frey M.

Finally, individual digits should be evaluated in the transverse and longitudinal planes using dynamic maneuvers as clinically indicated for the evaluation of pathology. There are five flexor tendon pulleys in the fingers, which are named A1–A5 and consist of annular ligament pulleys and cruciate pulleys—that is, the flexor tendon pulley system. The thumb only has two pulleys, which are labeled A1 and A2. When evaluating the pulley system in the digits, the A2 and A4 pulleys are most important in the sagittal plane, as shown in Figure 6-20. If pathology is present, this may demonstrate bowstringing and hypoechoic edema.[15]

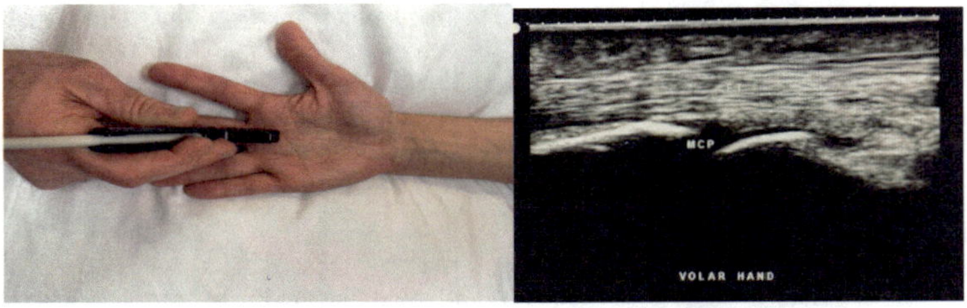

Figure 6-20: Side-by-side pictures of the transducer position and sagittal volar hand imaging of the metacarpal phalangeal joint (MCP) and the flexor tendon (FT). The abbreviations given here and labeled on the ultrasound image represent the corresponding structures.

The precision of ultrasound imaging and its relevance for carpal tunnel syndrome. Surg Radiol Anat. 2001;23(2):117–21. doi: 10.1007/s00276-001-0117-8. PMID: 11462859.

15. Msksono.org: Musculoskeletal Ultrasonography [internet]. [place unknown]; c2022 [cited 2023 Oct 28]. Available from: https://msksono.org/

6.8 Hip Sonography

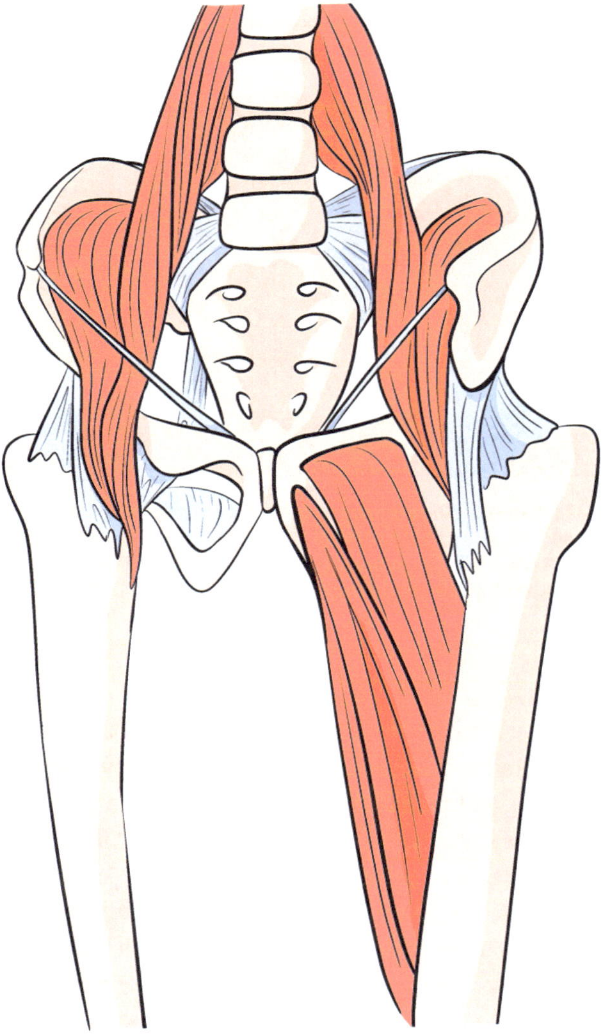

Figure 6-21: Anterior schematic of the skeletal structure and tendons of the hip.

Figure 6-21 shows a schematic of the skeletal structure and tendons of the hip. The low-frequency curvilinear transducer is most appropriate for hip evaluation. Start with the anterior evaluation by having the patient supine with the ipsilateral leg in full extension and with slight external rotation, as shown in Figure 6-22.

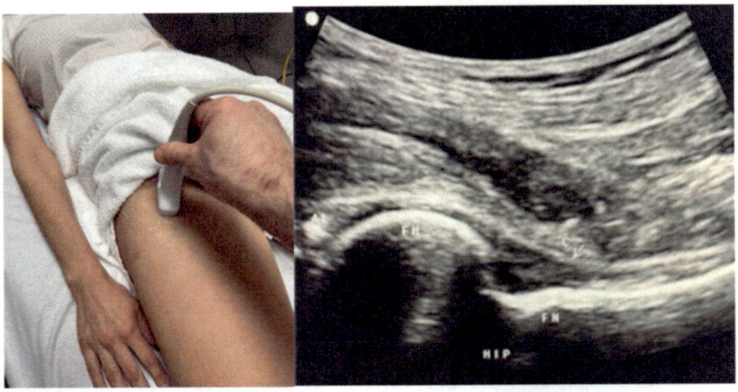

Figure 6-22: Side-by-side pictures of the transducer position and anterior hip sagittal imaging of the anterior labrum (AL), the femoral head (FH), the femoral neck (FN), and the femoral capsule (C). The abbreviations given here and labeled on the ultrasound image represent the corresponding structures.

Superficial to the capsule is the potential space between the capsule and the iliopsoas muscle, which is the iliopsoas bursa. This is the largest bursa in the human body, and an iliopsoas bursitis would be considered an extracapsular effusion. Like hip capsulitis, iliopsoas bursitis can be approached with an injection—but more superficial. The iliopsoas tendon is then evaluated by placing the transducer in the longitudinal plane in line with the femoral shaft and medial. The iliopsoas is a conjoined muscle composed of the iliacus and the psoas major muscles, which attach to the intertrochanteric line of the femur. This is evaluated from proximal to distal in the longitudinal and transverse planes. Also, consider evaluating the femoral vessels and nerve, sartorius muscle, tensor fascia lata tendons and muscles, lateral femoral cutaneous nerve, and rectus femoris tendon and muscles. Dynamic hip maneuvers may also help evaluate for tears, subluxation, or dislocation.[16]

Now have the patient in the lateral decubitus position with the hip to be evaluated up in a flexed 20–30 degree position to examine the gluteus muscles and tendons, as shown in Figures 6-23 through 6-25. The gluteus minimus, which is deep to the gluteus medius, originates from the ilium between the inferior and anterior gluteal lines. It inserts onto both the anterior aspect of the capsule and via its long head onto the anterior surface of the greater trochanter. The gluteus minimus and gluteus medius work together to abduct and internally rotate the hip. Finally, the gluteus maximus starts in the posterior iliac crest and sacrum/coccyx, crosses over the posterior facet, and inserts into the proximal femur, as Figure 6-25 shows. It extends and laterally rotates the hip.

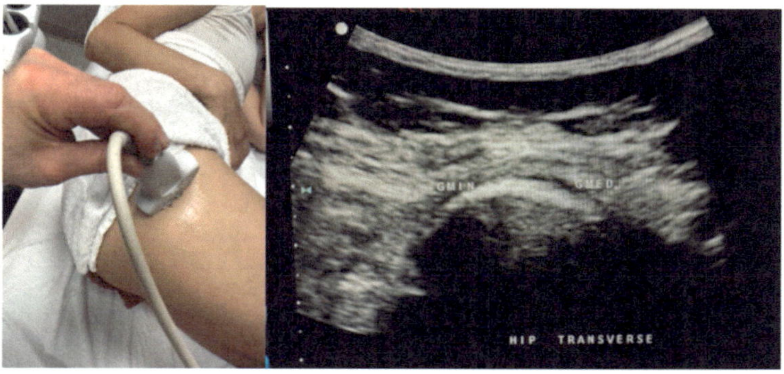

Figure 6-23: Side-by-side pictures of the transducer position and image of the gluteus minimus (GMIN) and gluteus medius (GMED) in the transverse plane. The abbreviations given here and labeled on the ultrasound image represent the corresponding structures.

16. Jacobson J, Kissin E, Lento P, Mazzola T, Moore RE, Shapiro S. Introduction to Musculoskeletal Ultrasound [DVD]. St Petersburg (FL): Gulfcoast Ultrasound Institute; 2014 Jan.

To visualize the tendons in the sagittal plane, rotate the probe 90 degrees and angle the beam anterior to posterior to visualize the gluteus minimus and posterior to anterior to visualize the gluteus medius, as shown in Figure 6-24, and the gluteus maximus, as shown in Figure 6-25.

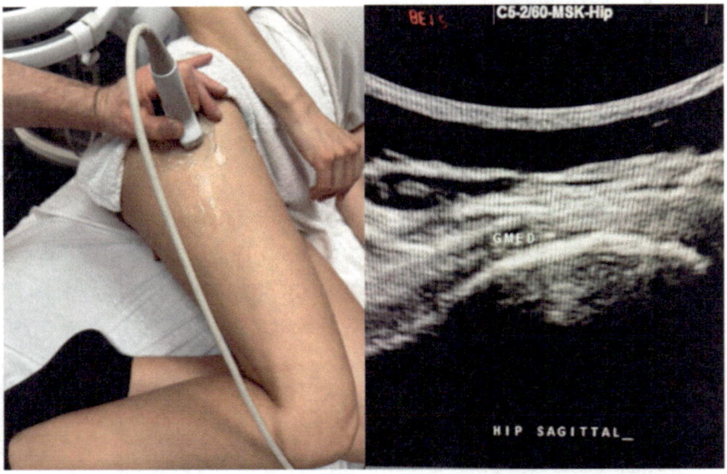

Figure 6-24: Side-by-side pictures of the transducer position and image of the gluteus medius (GMED) in the sagittal plane.

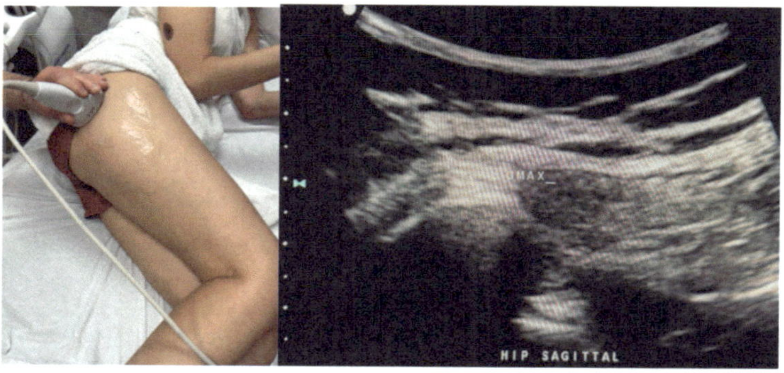

Figure 6-25: Side-by-side pictures of the transducer position and image of the gluteus maximus (GMAX) in the sagittal plane.

To evaluate the hamstrings, first identify the ischial tuberosity to locate the origins of the semimembranosus, biceps femoris, and semitendinosus tendons. Locate the conjoined tendons of the biceps femoris and semitendinosus, as shown in Figure 6-26. The semimembranosus lies deep and usually slightly inferior to the conjoined tendons.

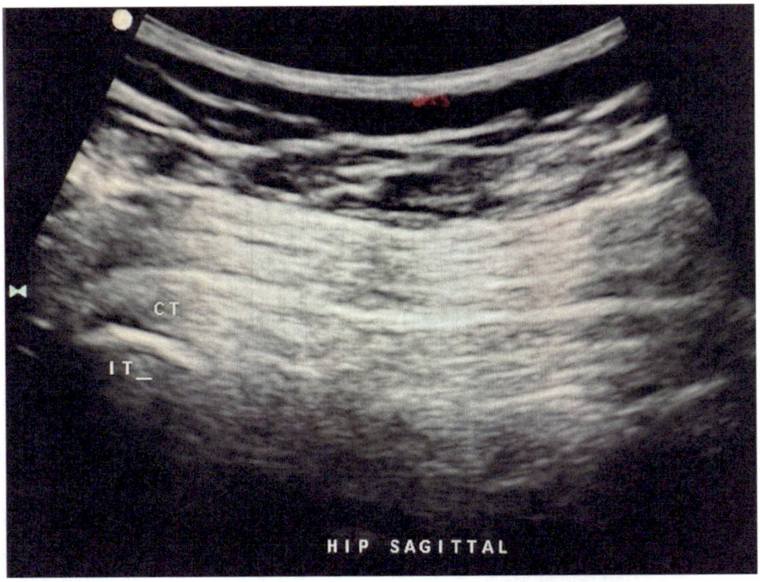

Figure 6-26: Sagittal hip image of the conjoined tendons (CT) of the biceps femoris and semitendinosus into the ischial tuberosity (IT). The abbreviations given here and labeled on the ultrasound image represent the corresponding structures.

6.9 Knee Sonography

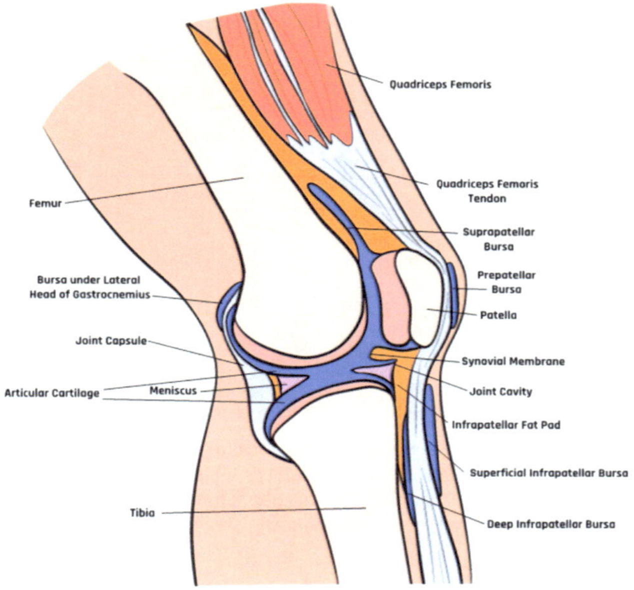

Figure 6-27: Medial schematic view of the knee anatomy.

Start the evaluation by using a high-frequency linear transducer and having the patient in the supine position. The anterior evaluation starts in the suprapatellar area with the probe in line with the femur, as shown in Figure 6-28.

Evaluate the following structures from deep to superficial, starting with the bony cortex of the femur, the quadriceps muscles and fascial planes, the femoral trochlea, the prefemoral fat pad, the suprapatellar bursa, and the suprapatellar fat pad. Evaluate proximally from the quadriceps muscle to the distal area over the patella in the longitudinal and transverse planes, looking for any pathology such as effusion or tears, as shown in Figures 6-29 and 6-30. It is sometimes helpful to perform toggling and heel-to-toe maneuvers to fine-tune the anatomy and avoid anisotropy.

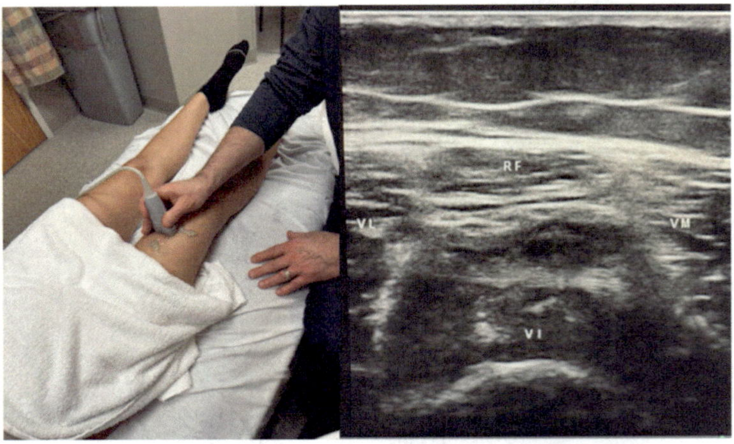

Figure 6-28: Side-by-side pictures of the transducer position and transverse image of rectus femoris (RF), vastus medialis (VM), vastus lateralis (VL), and vastus intermedius (VI). The abbreviations given here and labeled on the ultrasound image represent the corresponding structures.

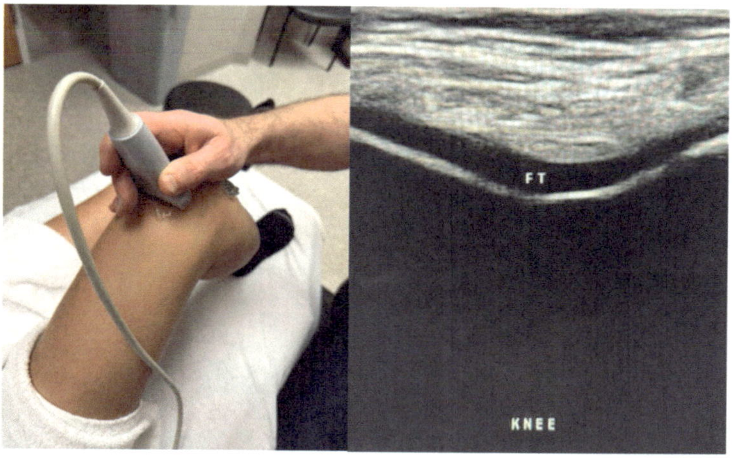

Figure 6-29: Side-by-side pictures of the transducer position and transverse image of the femoral trochlea (FT).

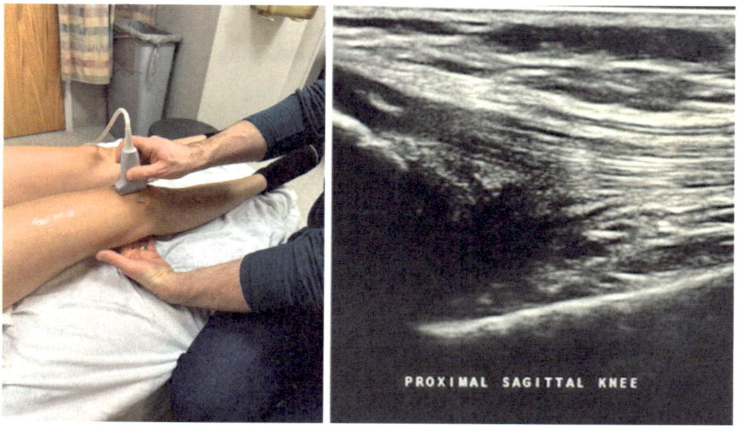

Figure 6-30: Side-by-side pictures of the transducer position and sagittal image of the proximal knee.

Next, slide the transducer medially, and evaluate the medial aspect of the knee joint from the femoral to the tibial condyles in the sagittal and transverse planes, as shown in Figures 6-31 and 6-32. Also, evaluate the medial collateral ligament and the medial meniscus in the joint space. As you move the transducer distally, evaluate the pes anserine complex for any evidence of injury or inflammation, such as pes anserine bursitis.

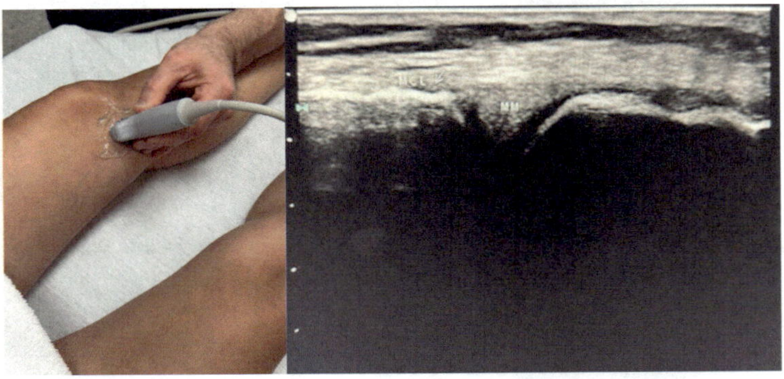

Figure 6-31: Side-by-side pictures of the ultrasound position and image of the medial meniscus (MM) and the medial collateral ligament (MCL). The abbreviations given here and labeled on the ultrasound image represent the corresponding structures.

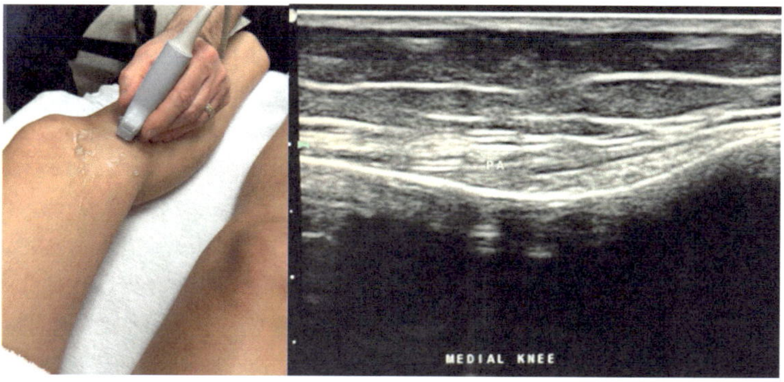

Figure 6-32: Side-by-side pictures of the transducer position and image of the pes anserine complex (PA).

Next, slide the probe to the lateral aspect of the knee joint, and evaluate the joint space of the distal femur and fibular head, as shown in Figure 6-33. In the longitudinal and transverse planes, evaluate the peripheral margin of the lateral meniscus and the lateral collateral ligament from proximal to distal.

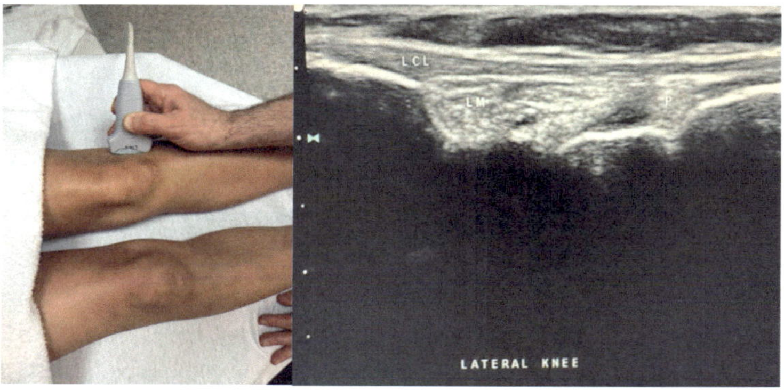

Figure 6-33: Side-by-side pictures of the transducer position and image of the lateral meniscus (LM), lateral collateral ligament (LCL), and popliteus (P). The abbreviations given here and labeled on the ultrasound image represent the corresponding structures.

Next, scan the infrapatellar area in the longitudinal plane with the bony landmarks proximally of the femur and tibia joint space and distally of the proximal tibia, as shown in Figure 6-34. Keep light pressure on the probe to avoid compressing any possible fluid within the bursa. There are two bursae superficial to the patellar tendon near the patella and one deep to the patellar ligament in the area of the proximal tibia. Evaluate the patellar ligament, sometimes called the patellar tendon, which is the portion of the quadriceps femoris tendon that continues from the patella to the tibial tuberosity.

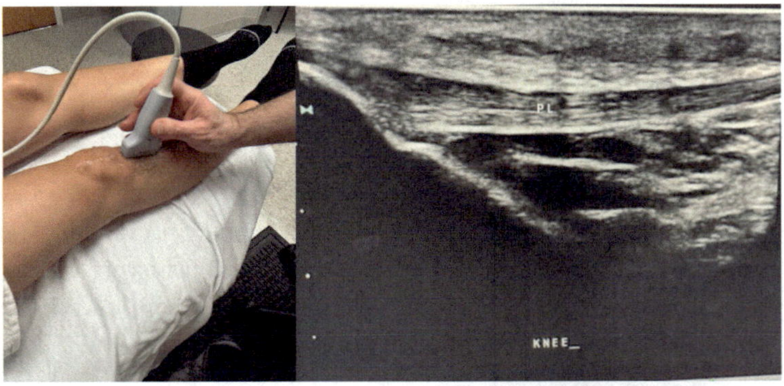

Figure 6-34: Side-by-side pictures of the sagittal image of the patellar ligament (PL).

Finally, the posterior view of the knee is evaluated with the knee slightly flexed at 10–20 degrees, as shown in Figure 6-35. Many structures can be seen in the popliteal fossa, including the popliteal artery and vein. One important area to evaluate is the area between the medial head of the gastrocnemius muscle and the semimembranosus tendon, which is the usual site of a Baker's cyst. This completes the knee evaluation.

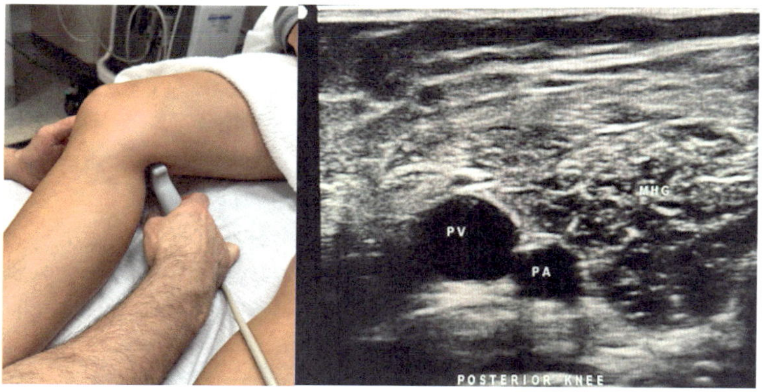

Figure 6-35: Side-by-side pictures of the transducer position and transverse image of the posterior knee viewing of the medial head of the gastrocnemius muscle (MHG), popliteal artery (PA), and the popliteal vein (PV). The abbreviations given here and labeled on the ultrasound image represent the corresponding structures.

6.10 Ankle and Foot Sonography

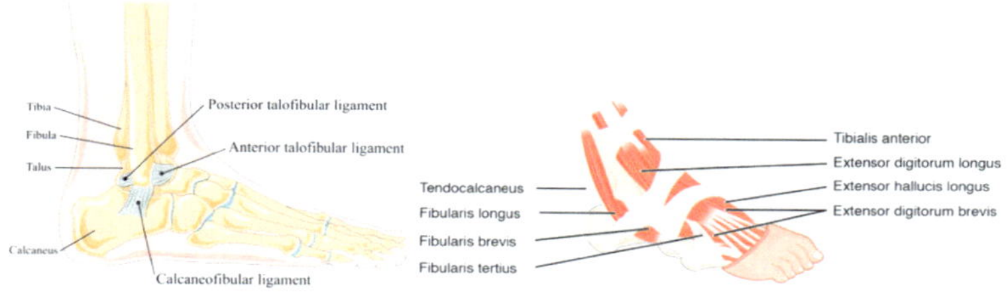

Figure 6-36: Anatomical view of the foot (Lateral collateral ligament of ankle joint by Laboratiores Servier licensed under CC BY-SA 4.0) and ankle (Dorsal superficial muscles of the right foot (lateral view) by Betts JG, Young kA, Wise JA, Johnson E, Poe B, Kruse DH, Korol O, Johnson JE, Womble M, and DeSaix P licensed under CC BY 4.0).

Figure 6-36 shows the anatomical view of the foot and ankle. The ankle and foot can be challenging to evaluate, since many structures require anatomic familiarity and detailed imaging. However, in general, a systematic approach is helpful. As with the other joint evaluations, a quadrant approach works best. Start with the anterior/dorsal evaluation by flattening the foot with an anterior longitudinal plane of the probe across the joint space of the tibia and talus, as shown in Figure 6-37. This provides an excellent focal point to sweep across the ankle joint to evaluate the muscles from medial to lateral: tibialis anterior, extensor hallucis longus, and extensor digitorum longus, as shown in Figure 6-38.

The subtalar joint is often of interest for evaluation and injection purposes. It is found in the longitudinal plane just medial to the lateral malleolus in the talus and calcaneus joint space.

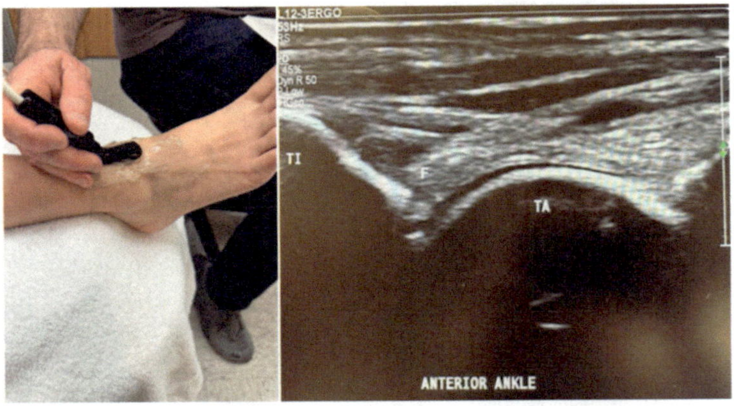

Figure 6-37: Side-by-side pictures of the transducer position and sagittal image of the anterior ankle, including the tibia (TI), talus (TA), and anterior fat pad (F). The abbreviations given here and labeled on the ultrasound image represent the corresponding structures.

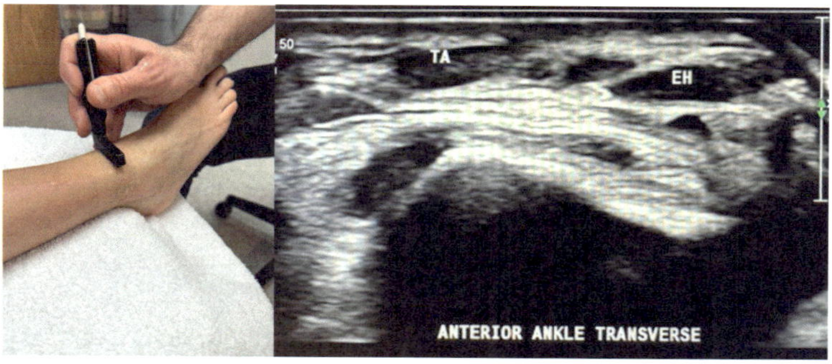

Figure 6-38: Side-by-side pictures of the transducer position and transverse image of the tibialis anterior (TA) and extensor hallucis (EH) in the ankle. The abbreviations given here and labeled on the ultrasound image represent the corresponding structures.

Evaluate from the proximal muscle to the tendon insertion points in the transverse and longitudinal planes as clinically indicated.[17] Dynamic imaging is helpful to evaluate the integrity of the ligament. Now place the probe behind the lateral malleolus in the longitudinal plane with a posterior to anterior angle of insonation to evaluate the peroneus longus, which is superficial to the peroneus brevis. Evaluate both the longitudinal and transverse planes proximally and distally to their insertion points, as shown in Figures 6-39 and 6-40.

17. Msksono.org: Musculoskeletal Ultrasonography [internet]. [place unknown]; c2022 [cited 2023 Oct 28]. Available from: https://msksono.org/

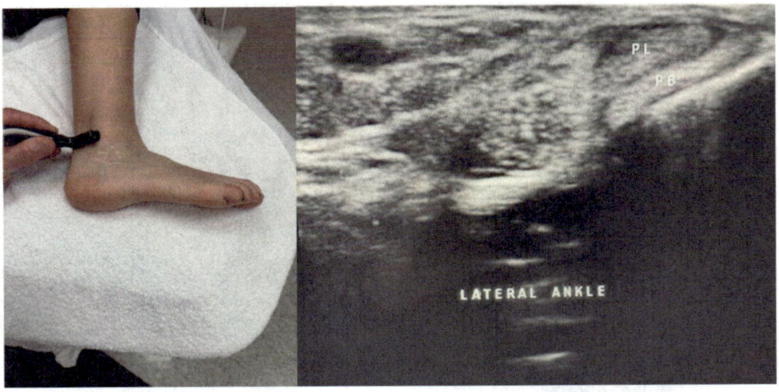

Figure 6-39: Side-by-side pictures of the transducer position and transverse image of peroneus brevus (PB) and peroneus longus (PL). The abbreviations given here and labeled on the ultrasound image represent the corresponding structures.

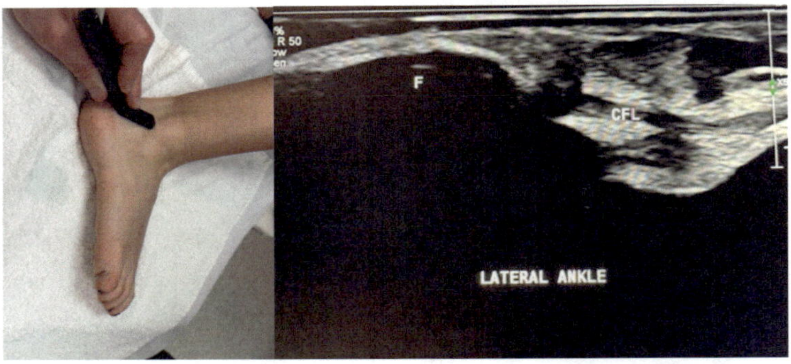

Figure 6-40: Side-by-side pictures of the transducer position and sagittal image of the calcaneofibular ligament (CFL), calcaneus (C), and fibula (F). The abbreviations given here and labeled on the ultrasound image represent the corresponding structures.

Place the probe in the transverse plane behind the medial malleolus to evaluate the medial side of the ankle and foot, as shown in Figure 6-41. Now we can evaluate the cross-sectional area of the structures in the tarsal tunnel. From medial to lateral, we have the **T**ibialis posterior tendon, flexor **D**igitorum longus tendon, posterior tibial **A**rtery/**V**ein/**N**erve, and flexor **H**allucis longus tendon. **T**om, **D**ick, **A**nd **V**ery **N**ervous **H**arry is a commonly used mnemonic to recall these anatomical structures. This is important to evaluate for tarsal tunnel syndrome if clinically indicated.[18]

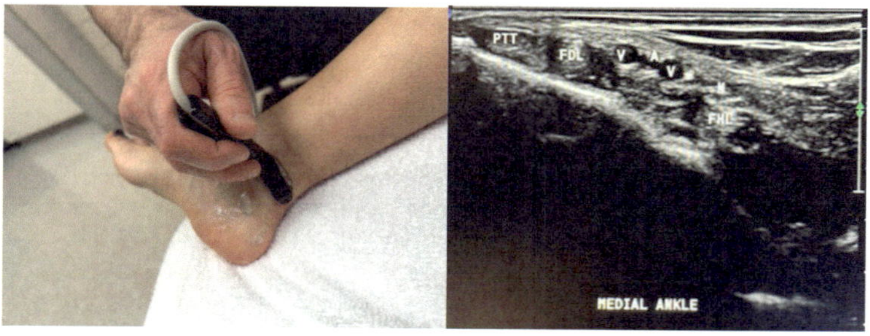

Figure 6-41: Side-by-side pictures of the transducer position and transverse image of the medial ankle, including the medial malleolus (MM), tibialis posterior tendon (PTT), flexor digitorum longus tendon (FDL), posterior tibial vein (V) and artery (A), tibial nerve (N), and flexor hallucis longus tendon (FHL). The abbreviations given here and labeled on the ultrasound image represent the corresponding structures.

To complete the evaluation, look at the Achilles tendon in the transverse and sagittal planes from the proximal gastrocnemius and soleus muscles to the insertion into the calcaneus, as shown in Figures 6-42 and 6-43.

18. Msksono.org: Musculoskeletal Ultrasonography [internet]. [place unknown]; c2022 [cited 2023 Oct 28]. Available from: https://msksono.org/

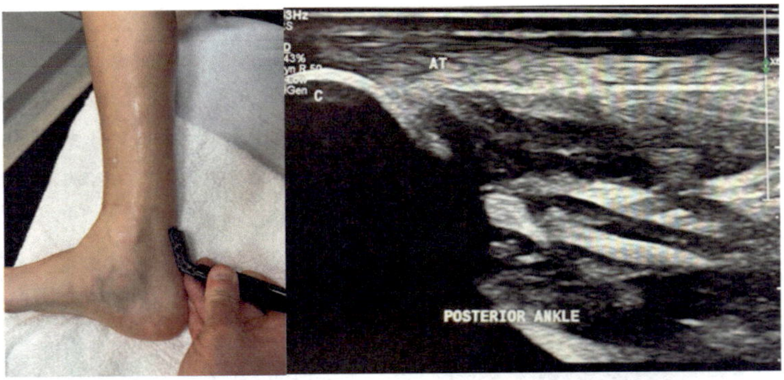

Figure 6-42: Side-by-side pictures of the transducer position and posterior ankle sagittal view of the Achilles tendon (AT) insertion into the calcaneus (C). The abbreviations given here and labeled on the ultrasound image represent the corresponding structures.

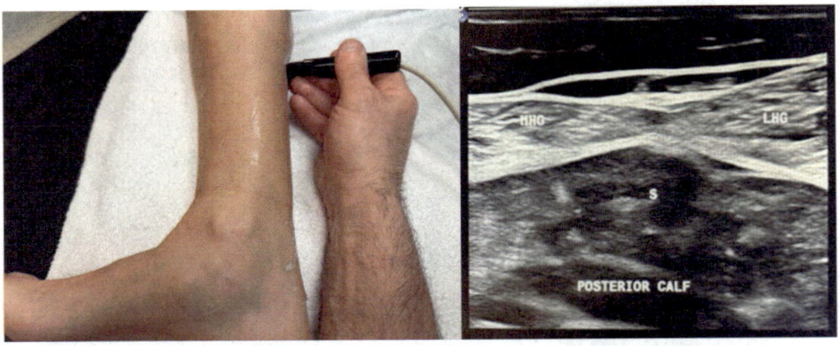

Figure 6-43: Side-by-side pictures of the transducer position and posterior calf transverse image include the medial head of the gastrocnemius muscle (MHG), the lateral head of the gastrocnemius muscle (LHG), and the soleus muscle (S). The abbreviations given here and labeled on the ultrasound image represent the corresponding structures.

Finally, evaluate the foot's plantar fascia in the longitudinal plane, as shown in Figure 6-44. The thickness at the insertion to the calcaneus should not be more than 4 mm, which would be suggestive of plantar fasciitis.

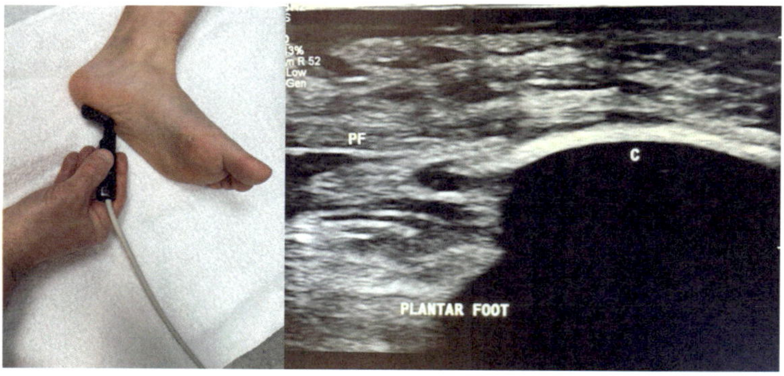

Figure 6-44: Side-by-side pictures of the transducer position and plantar sagittal view with plantar fascia (PF) insertion into the calcaneus (C). The abbreviations given here and labeled on the ultrasound image represent the corresponding structures.

6.11 Self-Assessment

1. What type of transducer is commonly used for most joint musculoskeletal evaluations?
2. What is the most common rotator cuff injury?
3. What is the best position to view common rotator cuff injuries?
4. Name a common elbow injury found with the MSK ultrasound.
5. What anatomical structures are best evaluated by a high-frequency hockey stick transducer in MSK

imaging?
6. Name a wrist syndrome that ultrasound can be used to help diagnose.
7. What type of transducer is most appropriate for hip evaluation?
8. What two structures would a Baker's cyst be found between?
9. What thickness should the standard plantar fascia not exceed at the insertion into the calcaneus?

6.12 Further Readings

1. Jacobson JA. Fundamentals of Musculoskeletal Ultrasound. 3rd ed. [place unknown]: Elsevier Saunders. 2017. 472 p.
2. Msksono.org: Musculoskeletal Ultrasonography [internet]. [place unknown]; c2022 [cited 2023 Oct 28]. Available from: https://msksono.org/
3. Bianchi S, Martinoli C. Ultrasound of the Musculoskeletal System. [place unknown]: Springer; 2007. 834 p.

7. The Heart and Echocardiography

7.1 Learning Objectives

After reviewing this chapter, you should be able to do the following:

1. Become familiar with some of the cardiac anatomical and physiological features.
2. Understand conditions related to the heart.
3. Examine ultrasound images from echocardiography.

7.2 Introduction

At various times in history, the heart has dominated much attention as the center of the function of living. Although no longer considered the center of emotions and thought, the heart continues its dominance in sonography in terms of both the frequency with which studies are done and the breadth of modalities used. Echocardiography is a rapidly growing and frequently used assessment tool in acute care settings.

7.3 Cardiac Anatomy and Physiology

The position of the heart in the thorax is much more variable than often expected. The heart may be in a more superior or lateral placement in the thoracic cavity due to abdominal anatomical abnormalities. Lung conditions, such as emphysema, may present a heart in a vertical orientation due to chronic lung overdistension. In most healthy individuals, the heart may be best seen in a view called the parasternal long axis, where the probe is held at the second or third intercostal space immediately left of the sternum, as shown in Figure 7-1. Directly below the palpated area will be the chamber walls of the left atrium and the left ventricle.

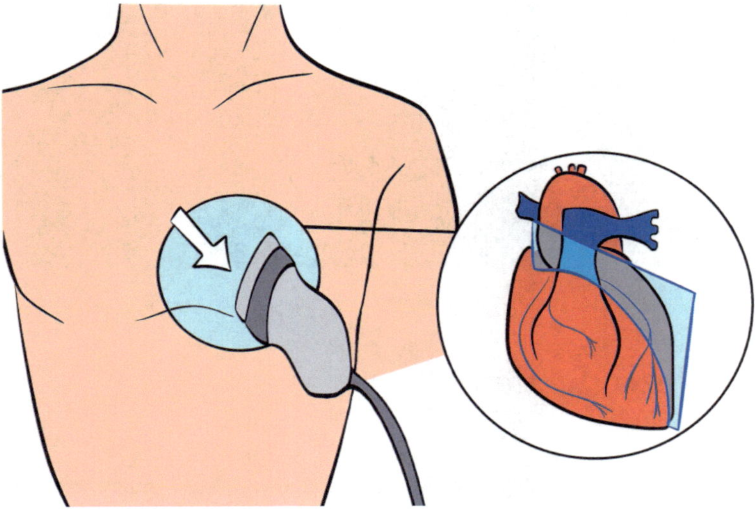

Figure 7-1: Parasternal long-axis view of the heart.

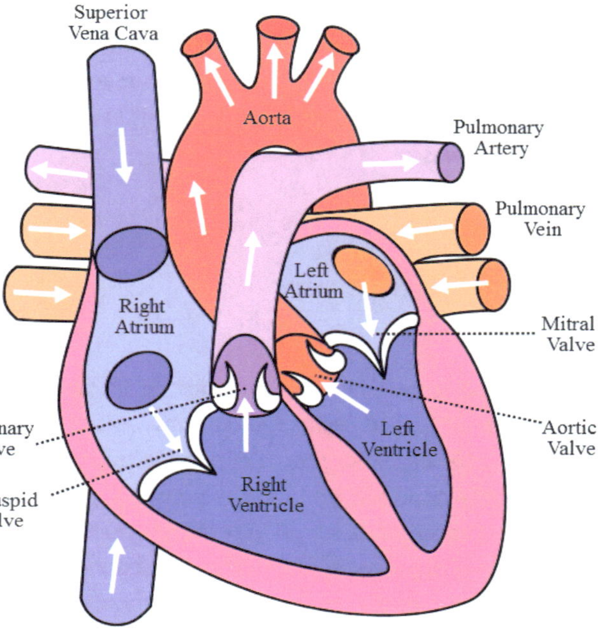

Figure 7-2: Blood circulation in the heart. Blood Circulation by Wapcaplet licensed under CC BY-SA 3.0

With optimal health, the heart contracts and expands synchronously. The circulation of the blood in the heart is shown in Figure 7-2. Blood enters the right atrium from the inferior vena cava and the superior vena cava. Blood then flows through the tricuspid valve to the right ventricle during cardiac expansion or diastole; this is mostly a passive flow. Blood is then expelled from the right ventricle during systole (or cardiac contraction) through the pulmonary artery and into the progressively smaller pulmonary arteries, pulmonary arterioles, and finally, pulmonary capillaries. Carbon dioxide and oxygen exchange occurs between the pulmonary capillaries and alveolar sacs of the lungs. Freshly oxygenated blood then returns via the pulmonary veins to the left atrium. Blood flows mostly passively from the left atrium, through the mitral valve, and into the left ventricle during diastole. Blood is finally propelled during systole through the aortic valve into the aorta and the body. The heart adapts to different behavioral and physiological stressors.

7.4 Echocardiography

A commonly performed, more sophisticated ultrasound evaluation of the heart is called *echocardiography*, often performed by technicians and cardiologists. Echocardiography machines are generally highly sophisticated (and expensive), with the capability and clarity of the views obtained improving over the years.

Some of the most sophisticated machines that are routinely used in echocardiography can evaluate wall motion of specific areas of the heart and perform static measurements of both the opening and closing of the valves that are cycling more frequently than once. Color flow Doppler features in these machines can simultaneously differentiate blood flows going through insufficient valves. These findings help determine what procedures are best for patients with heart disease. Some patient conditions and the ultrasound principles behind the evaluations are described below.

7.4.1 Ischemia

Ischemia is inadequate blood flow through the coronary arteries that may lead to myocardial (cardiac muscle cell) dysfunction or myocardial death. Ischemia typically results in chest pain. Temporary chest pain from reversible ischemia is called *angina*. Permanent myocardial tissue death from ischemia is called *myocardial infarction*. It is a common task of primary care providers to determine if the chest pain symptom an individual is experiencing is related to ischemia or another medical cause. Shortness of breath (dyspnea) is also a common symptom of ischemia. The goal is to address angina or reversible ischemia before irreversible global myocardial tissue death. The next step in addressing suspected coronary artery ischemia for patients fortunate enough to access the best health care is coronary artery catheterization. Further diagnosis and treatment with coronary artery catheterization are often done during the same procedure.[1] A coronary artery blockage can be treated with a small open mesh tube called a *stent*. If indicated, a stent or other procedure to reduce the blockage is completed to restore adequate blood flow.

It is common in the United States to try to diagnose those patients with risk factors and typical chest pain with an evaluation called a stress echocardiogram, also known as a "stress echo" or cardiopulmonary exercise testing with echocardiographic examination. In this test, the patient's heart rate is increased either by having them exercise on a treadmill or bike or by pharmacological measures to increase heart rate and heart contractility.[2] Pharmacological measures are used for patients with comorbidities, such as orthopedic or neurological conditions that may prevent safe treadmill or bike use. If there is a blockage of a single vessel in a stress echo, the part of the heart that this particular vessel supplies blood to will contract less efficiently. For

1. Chhabra L, Zain MA, Siddiqui WJ. Coronary Stents. [Updated 2023 Aug 7]. In: StatPearls [internet]. Treasure Island (FL): StatPearls Publishing; 2023 Jan–. Available from: https://www.ncbi.nlm.nih.gov/books/NBK507804/
2. Chhabra L, Zain MA, Siddiqui WJ. Coronary Stents. [Updated 2023 Aug 7]. In: StatPearls [internet]. Treasure Island (FL): StatPearls Publishing; 2023 Jan–. Available from: https://www.ncbi.nlm.nih.gov/books/NBK507804/

example, the left main coronary artery (LCA) supplies blood to the anterior part of the heart. In patients with a blockage of their LCA, the anterior portion of the heart will have noticeably less contractility. The reduction of contractility may be coupled with symptoms such as angina or dyspnea during physical movement. The contractility dysfunction that is induced by a rapid heart rate is often followed by cardiac catheterization. A coronary artery bypass grafting during an open-heart procedure may be needed in more severe cases. If so, a large incision is made in the chest to expose the heart. A vessel is removed from the patient's leg, chest, or arm and grafted around the coronary artery with blockage.

7.4.2 Myocardial Tissue Dysfunction

Myocardial muscle function has been assessed for over 200 years, and knowledge of congestive heart failure continues to change. Without imaging, it was hypothesized that heart contractility could be impaired and that the acceptance of incoming blood would be likewise impaired. This was realized to result in increased vascular pressure and fluid leakage in the peripheral interstitium, leading to ankle edema and edema in the lungs.[3] A stethoscope was used to hear the lung edema and probably first noted in an autopsy. Treatment at first involved trying to reduce intravascular volume by using a tourniquet or bloodletting. This was done with some success. One of the first medicines to treat it was a purification of products of the foxglove plant to produce a derivative of digitalis. To this day, medical providers continue to try to decrease intravascular

3. Malik A, Brito D, Vaqar S, et al. Congestive Heart Failure. [Updated 2022 Nov 7]. In: StatPearls [internet]. Treasure Island (FL): StatPearls Publishing; 2023 Jan–. Available from: https://www.ncbi.nlm.nih.gov/books/NBK430873/

volume and increase cardiac contractility. Digitalis derivatives like digoxin continue to be prescribed for abnormal heart rhythms.

Echocardiography is most important in the initial diagnosis and ongoing care of reduced myocardial contractility. Heart muscle contractility is assessed for quality and calculated as an ejection fraction (EF).[4] This is calculated by measuring the area of the left ventricle during diastole when the left ventricle is at the endpoint of receiving all of the blood prior to the contraction and at the point where the left ventricle has reached end-systole, which is the greatest point of contraction. The blood volumes corresponding to the endpoints during diastole and systole are referred to as end-diastolic volume (EDVol) and end-systolic volume (ESVol), respectively. Stroke volume (SVol) is calculated by subtracting ESVol from EDVol. EF can then be mathematically expressed as follows:[5]

$$EF = \frac{SVol}{EDVol} \times 100\% = \frac{EDVol - ESVol}{EDVol} \times 100\%.$$

If the left ventricle contains 100 ml of blood at the end of diastole and 40 ml when it fully contracts during systole, SVol is 60 ml, and thus, the EF is 60%. A normal EF percent is considered to be around 55–75%. EF percent measurements are becoming more reliable. This is largely due to software improvements allowing the area of a 3D object to be measured on a 2D screen. The description

4. Malik A, Brito D, Vaqar S, et al. Congestive Heart Failure. [Updated 2022 Nov 7]. In: StatPearls [internet]. Treasure Island (FL): StatPearls Publishing; 2023 Jan–. Available from: https://www.ncbi.nlm.nih.gov/books/NBK430873/
5. Kosaraju A, Goyal A, Grigorova Y, Makaryus AN. Left Ventricular Ejection Fraction. 2023 Apr 24. In: StatPearls [internet]. Treasure Island (FL): StatPearls Publishing; 2023 Jan–. PMID: 29083812.

of a patient having an EF percent of significantly less than 50% gives experienced clinicians a good idea of a patient's morbidity and mortality.

7.4.3 Valvular Dysfunction

Heart valves allow blood to travel from one area of the heart to another and prevent "backflow" from its new area to the old area. The four valves evaluated by echocardiography are the aortic valve (between the left ventricle and aorta), the mitral valve (between the left atrium and the left ventricle), the tricuspid valve (between the right atrium and right ventricle), and the pulmonary valve (between the right ventricle and the pulmonary artery). Clinically relevant dysfunction is much more common in the aortic and mitral valves. There are two common types of valvular dysfunction.

1. Narrowing of the valve (stenosis) allows an inadequate amount of flow through the aortic stenosis, which may cause both inadequate perfusion and backflow pressure to the left ventricle, resulting in abnormal left ventricular enlargement. Figure 7-3 shows a severe mitral valve stenosis on echocardiography.
2. Incompetence of the valve or valvular regurgitation allows the abnormal backflow of blood, causing a lack of perfusion and abnormal pressure into the contributing portion of the heart. Aortic regurgitation can be observed on echocardiograms, as shown in Figure 7-4.

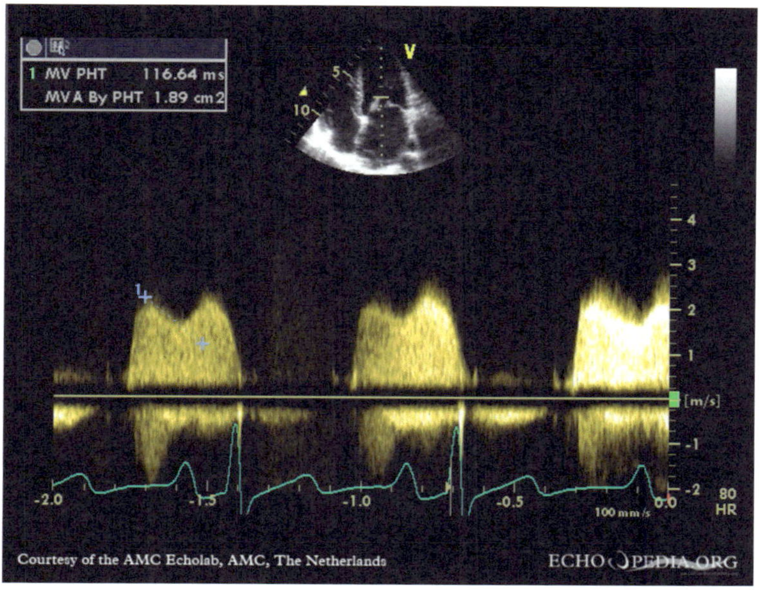

Figure 7-3: Echocardiogram showing severe mitral valve stenosis. Severe mitral valve stenosis E00615 (CardioNetworks ECHOpedia) by CardioNetworks licensed under CC BY-SA 3.0

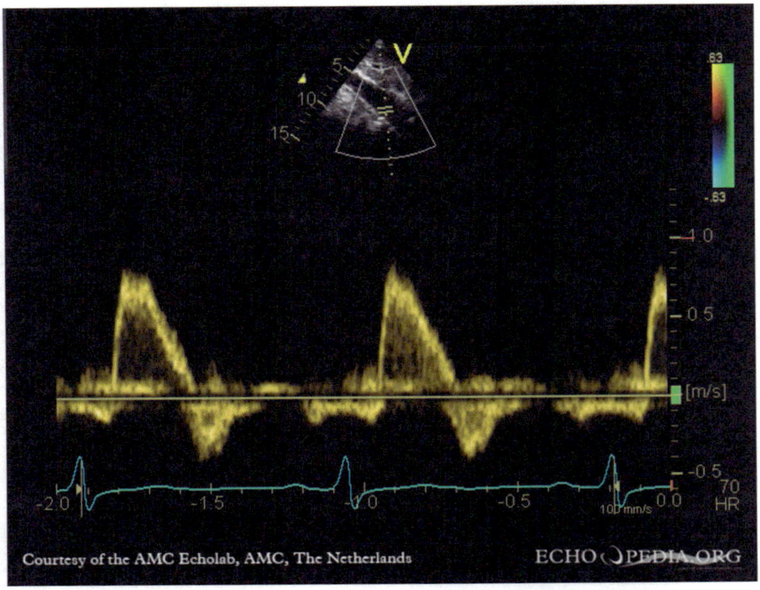

Figure 7-4: Echocardiogram showing dilated ascending aorta with severe aortic regurgitation. Dilated ascending aorta, severe aortic regurgitation E00449 (CardioNetworks ECHOpedia) by CardioNetworks licensed under CC BY-SA 3.0

Echocardiography evaluates valvular dysfunction by direct visualizations (measuring valve diameter in systole and diastole). This is obtaining static views for those changing most often more frequently than once per second in real time. Indirect measurements are obtained by measuring blood velocity through the valve and watching the color Doppler backflow of blood through valves with regurgitation.

7.5 Self-Assessment

1. What ultrasonic view gives the best access to the heart anatomy during cardiac ultrasound scanning for most healthy individuals?
2. What is ischemia? How does this lead to myocardial infarction?
3. How is the ejection fraction calculated?
4. What are the two common types of valvular dysfunction?

7.6 Further Readings

1. Singh S, Goyal A. The origin of echocardiography: A tribute to Inge Edler. Tex Heart Inst J. 2007;34(4):431–8. PMID: 18172524; PMCID: PMC2170493.
2. Hoffman JI, Kaplan S. The incidence of congenital heart disease. J Am Coll Cardiol. 2002 Jun 19;39(12):1890–900. doi: 10.1016/s0735-1097(02)01886-7. PMID: 12084585.
3. Kossaify A, Bassil E, Kossaify M. Stress Echocardiography: Concept and Criteria, Structure and Steps, Obstacles and Outcomes, Focused Update and Review. Cardiol Res. 2020 Apr;11(2):89–96. doi: 10.14740/cr851. Epub 2020 Mar 10. PMID: 32256915; PMCID: PMC7092766.

8. Pulmonary Ultrasound

8.1 Learning Objectives

After reviewing this chapter, you should be able to do the following:

1. Identify the challenges associated with pulmonary ultrasound.
2. Learn the most efficient and effective techniques used in pulmonary ultrasound.
3. Understand the expanded use of lung ultrasound in primary care, including COVID-19.

8.2 Introduction

As late as 1991, the most authoritative textbooks in internal medicine stated that pulmonary ultrasound was impossible due to the poor transmitting quality of air. Daniel Lichtenstein was the first to realize that all of the artifacts produced in the lungs represented different conditions, such as pulmonary edema, pneumothorax, pneumonia, and pulmonary effusion.[1] In recent years, various

1. Lichtenstein D, Mézière G, Biderman P, Gepner A, Barré O. The comet-tail artifact. An ultrasound sign of alveolar-interstitial

handheld ultrasounds have been extensively used for relatively quicker evaluations of lung conditions, including COVID-19, an infectious disease caused by the Severe Acute Respiratory Syndrome Coronavirus 2 (SARS-CoV-2) virus that was discovered in 2019.[2]

8.3 Cardiac Versus Pulmonary Ultrasound

While both systems are housed in the thorax, pulmonary and cardiac ultrasound approaches and findings can be strikingly different. As briefly discussed in the previous chapter, cardiac ultrasound has specific measures used for valves, blood velocity, and cardiac muscle contractility. The operators' abilities can adversely affect measurement accuracy on such subtle issues as the angle at which the probe is held and, subsequently, the angle at which measurements are obtained. One of the productive uses of artificial intelligence in the ultrasound field allows the operator to know when the best angle of insonation is obtained before the image is captured. Clear imaging is partially achieved by filtering raw images and removing portions of the images that are confusing to obtain the measurements. These measurements then determine important treatment routes, including choosing medication for the patient and even deciding if cardiac surgery is necessary.

syndrome. Am J Respir Crit Care Med. 1997 Nov;156(5):1640–6. doi: 10.1164/ajrccm.156.5.96-07096. PMID: 9372688.

2. Cascella M, Rajnik M, Aleem A, et al. Features, Evaluation, and Treatment of Coronavirus (COVID-19) [Updated 2023 Aug 18]. In: StatPearls [internet]. Treasure Island (FL): StatPearls Publishing; 2023 Jan–. Available from: https://www.ncbi.nlm.nih.gov/books/NBK554776/

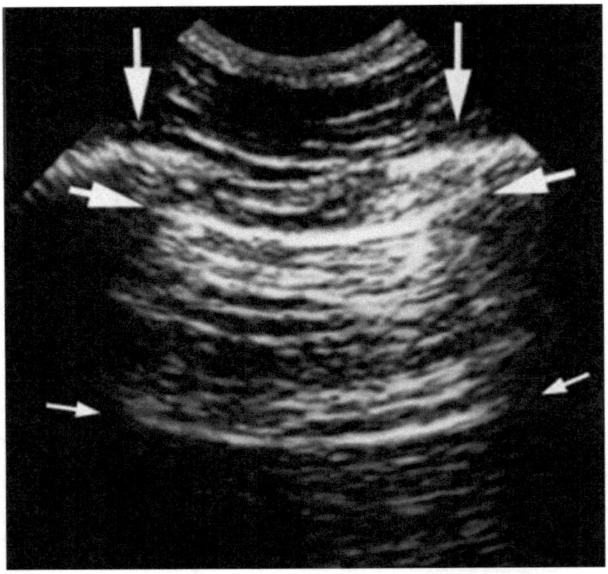

Figure 8-1: Ultrasound image of the intercostal space with the ribs shown by the vertical arrows. Rib shadows are displayed below. The upper horizontal arrows represent the pleural line, and the lower horizontal arrows represent the artifact of the pleural line, called the A-line. Normal lung surface by Daniel A. Lichtenstein licensed under CC BY 2.0

In many ways, pulmonary ultrasound is the antithesis of cardiac ultrasound. The motion cycle of the lungs is less dramatic and less frequent, and the "snow patterns" on the screen caused by a lung consolidation or other conditions allow health care providers to make rapid, critical, and often lifesaving decisions about patients who are most ill. The total picture of the reflections of an ultrasound beam that is received from the same beam-emitting probe is often blurred. The "artifacts" are often distracting and even prohibitive in making the exact measurements. Figure 8-1 shows an ultrasound image of the intercostal space. Daniel Lichtenstein, known by many as the father of critical care ultrasound, was one of the first experts to recognize that the artifacts emitted in the lungs were not the problem but part of the solution in determining the pulmonary and vascular status of the patient.[3] The notion of lung ultrasound was possible and pertinent through Dr. Lichtenstein and his colleagues' recognition and hard work.

8.4 A Brief Pathophysiology of the Lungs

Successful respiratory cycles are controlled at many levels, from the brain to the alveoli. Clinically, we focus on what we can change to improve our patients' conditions. The brain's complex mechanisms and primary circuits, such as the autonomic nervous system and vagus nerve, are often out of our reach to meaningfully control for the long-term. We can mimic respiratory control by completely paralyzing our muscles and forcing air with oxygen concentrations into the lungs at a rate determined to be optimal by the clinician. This action is almost always temporary, such as for patients needing general anesthesia for surgery or patients in respiratory failure needing intensive care unit (ICU) attention. Lung ultrasound is most often used to help diagnose the cause of acute respiratory failure or evaluate continuing critical care efforts on patients already being actively treated for respiratory failure.

3. Lichtenstein D, Mézière G, Biderman P, Gepner A, Barré O. The comet-tail artifact. An ultrasound sign of alveolar-interstitial syndrome. Am J Respir Crit Care Med. 1997 Nov;156(5):1640–6. doi: 10.1164/ajrccm.156.5.96-07096. PMID: 9372688.

8.5 Ultrasound Assessment of Some Lung Conditions

Low-frequency curvilinear transducers with frequencies in the range of 3–6 MHz are best suited for lung ultrasound. Much of the recorded action of the lung begins and ends with the pleura. The pleura, a general term for the parietal pleura, air, and visceral pleura, is the denser outer covering of the lung that moves with the lung. From a sonographer's perspective, the pleura is the most easily visualized lung structure. Inspecting the pleura sonographically allows for observations of both pleural motion and pleural sonographic disruption.

8.5.1 Disruptions of Pleural Motion

As the pleura moves, it moves the rest of the lung. This "sliding lung sign" accurately indicates lung motion and successful ventilation. If the pleura is not sliding, the lung, at least on the side of the thorax being inspected, is not ventilating thoroughly. Quick speculations about the causes of a lack of ventilation can be confirmed and treated to save lives. Below are some causes of a lack of ventilation that can be considered when there is no sliding lung sign:[4]

1. Pneumothorax
2. Adhesions of the surface of the pleura from infection or autoimmune disease
3. Pleural effusions not allowing lung expansion

4. Lichtenstein DA. Lung ultrasound in the critically ill. An Intensive Care. 2014 Jan 9;4(1):1. doi: 10.1186/2110-5820-4-1. PMID: 24401163; PMCID: PMC3895677.

4. Central or peripheral nervous system malfunction by not stimulating proper inspiration
5. Neuromuscular failure impacting the diaphragm or intercostal muscles that influence adequate inspiration

8.5.2 Disruptions of Pleural Consistency

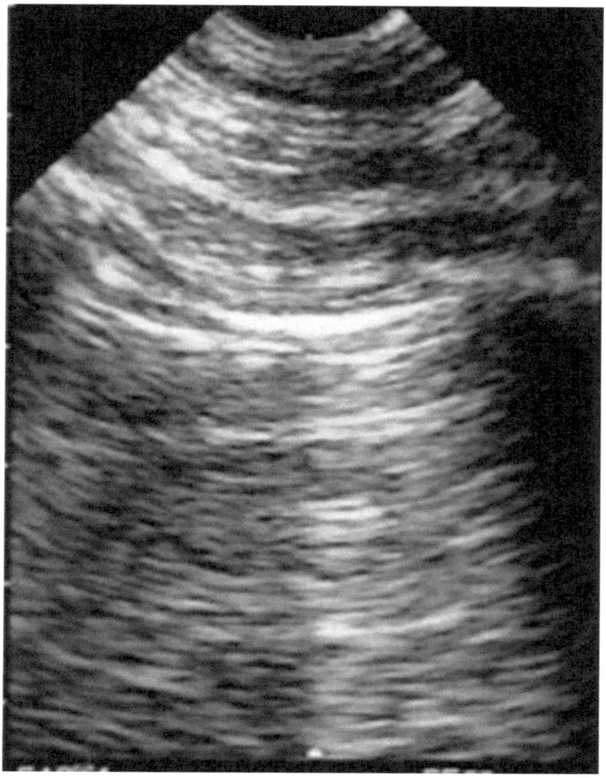

Figure 8-2: Pleural line with A-lines similar to Figure 8-1, indicating gas below the pleural line. Pneumothorax and the stratosphere sign by Daniel A. Lichtenstein licensed under CC BY 2.0

What have been named "A-lines" are the results of a reverberation artifact from the pleura or recurrent mirror images of the pleura that appear under the pleural lining at regular intervals, as shown in Figure 8-2. Disruption of the interstitium of the pleura changes the entire pleural reflection. Because the beam is prevented from uniformly penetrating the pleura, the areas that are disrupted are now seen as "B-lines," as shown in Figure 8-3. There is no longer a reverberation artifact or A-lines when there are B-lines in motion.

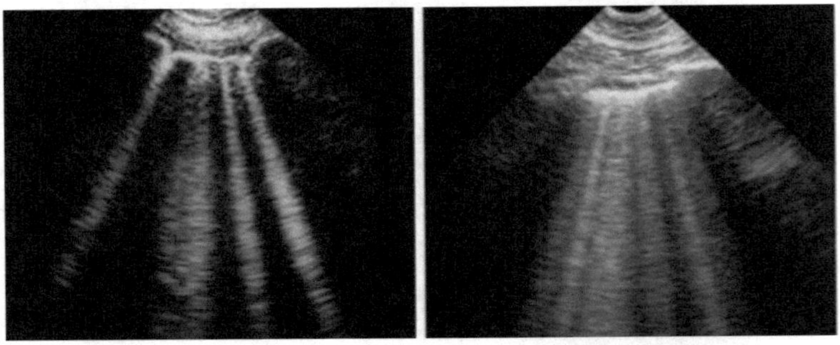

Figure 8-3: The image on the left shows four or five B-lines. The image on the right shows twice as many B-lines with two examples of pulmonary edema. Interstitial syndrome and the lung rockets by Daniel A. Lichtenstein licensed under CC BY 2.0

B-lines representing pleural interstitial disruption often represent pulmonary edema or vascular overload, as shown in the right diagram of Figure 8-3. The edema, which collects on the pleural surface, causes "comet tail" artifacts that account for the lines extending the lungs' entire length. It is often in the realm of intensivists or pulmonologists to ascertain a patient's intravascular volume. Dr. Lichtenstein was an original contributor to the direct visualization of the inferior vena cava to determine intravascular blood volume. He later discovered that B-lines might more accurately represent intravascular volume overload.

8.5.3 Lung Parenchyma

Ultrasound principles allow us to understand that the complete reflection of the waves due to necrotic debris is represented as a bright image on the screen in the area of the lung parenchyma. Nonsonographic clues such as a fever may lead the clinician to speculate that the bright reflection is an infection of the alveolar cells or pneumonia. Since alveoli have such thin walls that allow oxygen to be exchanged, it is often accurately feared that this same thin wall would easily allow infectious organisms to also cross from the alveoli to the bloodstream. Noninfectious obstructions around the alveoli appear as a similar bright reflection of the ultrasound image. This scarring is known as *atelectasis* and is often considered much less dangerous to the patient.

8.6 The Reciprocal Development of COVID-19 and Its Assessment

Due to many factors, it is a slippery slope to write anything about COVID-19. The disease is widespread, the actual effectiveness of isolation is unknown, testing has been inaccurate for the most part, supplies to even care for patients have been limited to an unprecedented extent in most developed countries, and the treatment theories change almost daily.

Far too many patients diagnosed with COVID-19 have had a rapid downhill course and go on to have a painfully long critical care course with a high mortality rate. For the first time in the world, there have been several considerations brought to the forefront. There has been a common consensus that COVID-19 has caused significant shortages of medical supplies both in the United States and worldwide, including those in testing capacity, ICU and hospital bed supply, hospital staff, personal protective equipment (PPE), and mechanical ventilators for affected regions.[5] There is greatly magnified attention on the safety of caretakers of the most critically ill patients. The mental health and well-being of health care professionals have been the focus of increased attention, with persistent evidence of high burnout, psychosocial stress, and mortality rates.[6] There has been a realization of how the increased vulnerability of our underserved population will perhaps impact how medical care is distributed in the future. The prevalence of COVID-19 has had a disproportionate impact on the poor, minorities, and a broad range of vulnerable populations due to its inequitable spread in areas of dense population and limited mitigation capacity resulting from a high prevalence of chronic conditions or poor access to high-quality public health and medical care.[7] There has been a growing need for more research on health equity in order to increase global knowledge and allow cross-national learning of what works for those most in need due to the direct and collateral effects of COVID-19. It has also been demonstrated that a pandemic can quickly destroy many

5. Dar M, Swamy L, Gavin D, Theodore A. Mechanical-Ventilation Supply and Options for the COVID-19 Pandemic. Leveraging All Available Resources for a Limited Resource in a Crisis. Ann Am Thorac Soc. 2021 Mar;18(3):408–416. doi: 10.1513/AnnalsATS.202004-317CME. PMID: 33202144; PMCID: PMC7919160.
6. Gupta N, Dhamija S, Patil J, Chaudhari B. Impact of COVID-19 pandemic on healthcare workers. Ind Psychiatry J. 2021 Oct;30(Suppl 1):S282–S284. doi: 10.4103/0972-6748.328830. Epub 2021 Oct 22. PMID: 34908710; PMCID: PMC8611576.
7. Shadmi E, Chen Y, Dourado I, Faran-Perach I, Furler J, Hangoma P, Hanvoravongchai P, Obando C, Petrosyan V, Rao KD, Ruano AL, Shi L, de Souza LE, Spitzer-Shohat S, Sturgiss E, Suphanchaimat R, Uribe MV, Willems S. Health equity and COVID-19: Global perspectives. Int J Equity Health. 2020 Jun 26;19(1):104. doi: 10.1186/s12939-020-01218-z. PMID: 32586388; PMCID: PMC7316580.

characteristics of our society, affecting the economic, social, and personal habits of people in all countries. The World Health Organization has warned of a mental health burden related to the spread of COVID-19 infection through the global population: stress, worry, fear, and changes in our daily lives (working from home, temporary unemployment, homeschooling, etc.) are all challenging people's mental and physical health as well as the global health care system and economy.[8]

COVID-19 has been a very humbling condition to deal with for any health care professional. Many patients do well even with a positive COVID test. Many patients found to be COVID-positive on routine screening remain asymptomatic. On the other end of the care spectrum, too many people without major risk factors are experiencing significant acute and chronic symptoms from the virus. Many have died, and a handful of them have been health care workers. The objective predictors of who will do well, such as lab tests and plain X-rays, have been inaccurate in too many cases. Especially in developed countries, diagnosticians can usually rely on numerous pieces of data to arrive at a reasonable treatment plan. At the time of this writing, COVID-19 remains a disease where the outcome can be unpredictable in a most painful manner. Patients with comorbidities are indeed at far greater risk of doing poorly, up to the point of having a much higher mortality rate. Age, congestive heart failure, chronic obstructive pulmonary disease, and dementia significantly increase the chance of doing poorly. The realization that an extensive ICU run for patients with limited life expectancy may not be in the best interest of the patient or a society where

8. World Health Organization (WHO). Mental Health and COVID-19: Early evidence of the pandemic's impact. [place unknown]: Mental Health and Substance Use, WHO Headquarters; 2022 Mar 2. 13 p. Report No.: WHO/2019-nCoV/Sci_Brief/Mental_health/2022.1. Available from: https://www.who.int/publications/i/item/WHO-2019-nCoV-Sci_Brief-Mental_health-2022.1

ICU resources are limited has been a most painful reality for medical workers to consider.

Lung ultrasound with handheld machines has developed into the most useful serial evaluator of the progression of COVID-19 lung disease. One of the challenges during the COVID-19 pandemic is was that several patients had to undergo ultrasound examinations within a limited amount of time, and those ultrasound tests needed to be completed on easily transportable machines. It was imperative for these machines to have the least number of knobs and small unreachable parts or spaces so that they can be sterilized quickly for immediate use with other COVID-19 patients.

Returning to the earlier chapters, we recall that degrees of brightness represent specific tissue characteristics, such as density. Pleura is one of those highly reflective densities. As the acute phase of COVID-19 disease progresses, distinctive changes may be observed first on lung ultrasound, which warns the clinician that the patient may be quickly worsening. Initially, COVID-19 may begin to change the pleura and cause the patient's condition to worsen due to pleural thickening and focal fluid collection in some areas. Pleural thickening can be relatively easily seen on lung ultrasound, as shown in Figure 8-4.

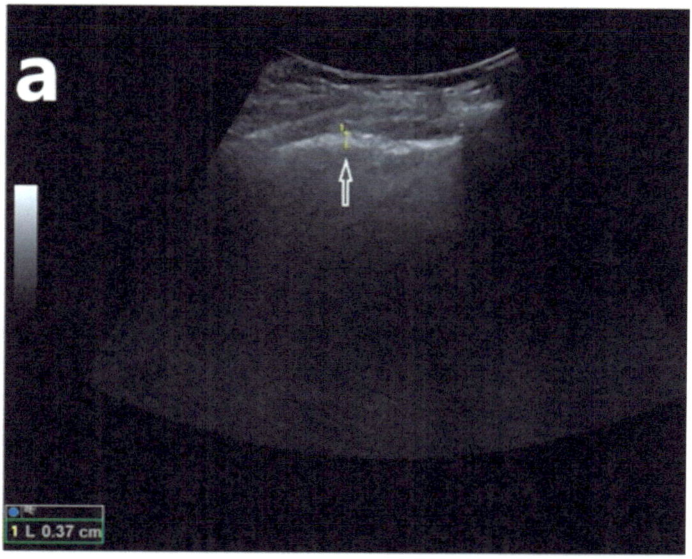

Figure 8-4: B-line artifacts arising from an apparently thickened pleural line. B-line artifacts by Buda N, Cylwik J, Mróz K, Rudzińska R, Dubik P, Malczewska A, Oraczewska A, Skoczyński S, Suska A, Górecki T, Mendrala K, Piotrkowski J, Gola W, Segura-Grau E, Zamojska A, and Wełnicki M licensed under CC BY 4.0

As the thickened pleura from COVID-19 becomes even more inflamed, adhesions can develop in specific areas of the lungs. These adhesions may also be focal so that there may be focal areas of impaired ventilation even on the same side of the lung. As COVID-19 disease progresses, there can be focal infiltrates in the posterior and inferior portions of the lung, which are unique and can be identified on an ultrasound as gray areas (these regions appear anechoic on the normal lungs) commonly referred to as "ground glass". The nature of the infiltrates with "ground glass" is present. The "ground glass" is from inflammation of the bronchioles and alveoli. This type of inflammation is characteristic of COVID-19.

It has been suggested that the ultrasound or CT findings of

infiltrates in certain distributions can also be used as a diagnostic test suggesting the need for imploring further specific COVID-19-influenced testing. CT almost always presents a more precise view of the anatomic effects of the COVID virus on a patient. Handheld ultrasound is more accurate in evaluating the physiological effects of COVID-19 on ventilation, especially when considering the sliding lung sign. CT is more expensive and delivers more radiation. It may not be as practical to clean thoroughly to prevent droplet transmission of infection.

As opposed to other serious systemic diseases, progressive lung findings may remain focal or become more diffuse in the lungs. As the disease progresses, a more diffuse systemic condition from adhesions and inflammation may develop. This is known as *acute respiratory distress syndrome* (ARDS), which often leads to the need for ventilator support for several weeks. Several characteristics of COVID-19 have forced ethical discussions of patient care to very different levels from those for patients with other conditions.

Given the dismal results of cardiopulmonary resuscitation (CPR) in treating patients with COVID and the genuine risk to presumably healthy medical professionals, modifications to CPR have been suggested by many institutions that influence health care actions. These include the following:[9]

1. Do not begin CPR if there is inadequate PPE for the CPR caretakers or a lack of training in the use of PPE.
2. The number of caretakers in a room where a patient with

9. Nolan JP, Monsieurs KG, Bossaert L, Böttiger BW, Greif R, Lott C, Madar J, Olasveengen TM, Roehr CC, Semeraro F, Soar J, Van de Voorde P, Zideman DA, Perkins GD; European Resuscitation Council COVID-Guideline Writing Groups. European Resuscitation Council COVID-19 guidelines executive summary. Resuscitation. 2020 Aug;153:45–55. doi: 10.1016/j.resuscitation.2020.06.001. Epub 2020 Jun 7. PMID: 32525022; PMCID: PMC7276132.

COVID-19 is undergoing CPR should be limited to active caretakers only. This is to be done even at the expense of education for the less experienced.
3. CPR should be modified to not manipulate the respiratory system unless there is specialized equipment to limit COVID-19 spread. Consider compression-only CPR if bag-mask ventilation is difficult. Consider a mechanical compression device if there is a need for prolonged CPR.
4. Open discussions should be had with patients who would have a low chance of out-of-hospital survival if there were cardiopulmonary arrest or other emergent conditions. It is a solemn consideration even to begin CPR on a patient with COVID-19 when there is cardiopulmonary arrest, and the chances of meaningful survival of the patient are low.

Some of the humbling parts of COVID-19 can be seen in the personal experiences of clinicians who have authentic discussions with patients with COVID-19 and the statistics of their success or failure in helping the most ill:

1. Hasan et al.[10] systematically reviewed mortality in COVID-19 patients with ARDS. According to their work published in September 2020, they found that the overall pooled mortality estimate among 10,815 ARDS cases in COVID-19 patients was 39%.
2. In a study conducted by Bielski et al.[11] in November 2021, there

10. Hasan SS, Capstick T, Ahmed R, Kow CS, Mazhar F, Merchant HA, Zaidi STR. Mortality in COVID-19 patients with acute respiratory distress syndrome and corticosteroids use: A systematic review and meta-analysis. Expert Rev Respir Med. 2020 Nov;14(11):1149–1163. doi: 10.1080/17476348.2020.1804365. Epub 2020 Sep 29. PMID: 32734777; PMCID: PMC7544968.
11. Bielski K, Szarpak A, Jaguszewski MJ, Kopiec T, Smereka J, Gasecka

was no significant difference between pre-COVID-19 and COVID-19 periods in bystander-witnessed arrests, bystander CPR, and the use of mechanical chest compression. However, since the COVID-19 pandemic began, cardiac arrest was observed to be more frequent at home, and bystanders used automated external defibrillators less frequently.

3. A 2010 review of 79 studies[12] involving 142,740 patients found that the survival rate of out-of-hospital cardiac arrest was only 7.6%. Bystander-initiated CPR may increase those odds to 10%. Survival after CPR for in-hospital cardiac arrest was slightly better[13] but still only about 17%.
4. Although experience with COVID-19 continues to grow, reported mortality rates range from 50% to 97% in those requiring mechanical ventilation.[14],[15] These are significantly

A, Wolak P, Nowak-Starz G, Chmielewski J, Rafique Z, Peacock FW, Szarpak L. The Influence of COVID-19 on Out-Hospital Cardiac Arrest Survival Outcomes: An Updated Systematic Review and Meta-Analysis. J Clin Med. 2021 Nov 27;10(23):5573. doi: 10.3390/jcm10235573. PMID: 34884289; PMCID: PMC8658174.

12. Sasson C, Rogers MA, Dahl J, Kellermann AL. Predictors of survival from out-of-hospital cardiac arrest: A systematic review and meta-analysis. Circ Cardiovasc Qual Outcomes. 2010 Jan;3(1):63–81. doi: 10.1161/CIRCOUTCOMES.109.889576. Epub 2009 Nov 10. PMID: 20123673.
13. Girotra S, Nallamothu BK, Spertus JA, Li Y, Krumholz HM, Chan PS; American Heart Association Get with the Guidelines–Resuscitation Investigators. Trends in survival after in-hospital cardiac arrest. N Engl J Med. 2012 Nov 15;367(20):1912–20. doi: 10.1056/NEJMoa1109148. PMID: 23150959; PMCID: PMC3517894.
14. Richardson S, Hirsch JS, Narasimhan M, Crawford JM, McGinn T, Davidson KW; the Northwell COVID-19 Research Consortium; Barnaby DP, Becker LB, Chelico JD, Cohen SL, Cookingham J, Coppa K, Diefenbach MA, Dominello AJ, Duer-Hefele J, Falzon L, Gitlin J,

higher than the published mortality rates ranging from 35% to 46% for patients intubated with H1N1 influenza pneumonia and other causes of ARDS.[16],[17] Mortality was significantly associated with older age, lower body mass index, chronic

Hajizadeh N, Harvin TG, Hirschwerk DA, Kim EJ, Kozel ZM, Marrast LM, Mogavero JN, Osorio GA, Qiu M, Zanos TP. Presenting Characteristics, Comorbidities, and Outcomes Among 5700 Patients Hospitalized With COVID-19 in the New York City Area. JAMA. 2020 May 26;323(20):2052–2059. doi: 10.1001/jama.2020.6775. Erratum in: JAMA. 2020 May 26;323(20):2098. PMID: 32320003; PMCID: PMC7177629.

15. Arentz M, Yim E, Klaff L, Lokhandwala S, Riedo FX, Chong M, Lee M. Characteristics and Outcomes of 21 Critically Ill Patients With COVID-19 in Washington State. JAMA. 2020 Apr 28;323(16):1612–1614. doi: 10.1001/jama.2020.4326. PMID: 32191259; PMCID: PMC7082763.
16. Bellani G, Laffey JG, Pham T, Fan E, Brochard L, Esteban A, Gattinoni L, van Haren F, Larsson A, McAuley DF, Ranieri M, Rubenfeld G, Thompson BT, Wrigge H, Slutsky AS, Pesenti A; LUNG SAFE Investigators; ESICM Trials Group. Epidemiology, Patterns of Care, and Mortality for Patients With Acute Respiratory Distress Syndrome in Intensive Care Units in 50 Countries. JAMA. 2016 Feb 23;315(8):788–800. doi: 10.1001/jama.2016.0291. Erratum in: JAMA. 2016 Jul 19;316(3):350. Erratum in: JAMA. 2016 Jul 19;316(3):350. PMID: 26903337.
17. Estenssoro E, Ríos FG, Apezteguía C, Reina R, Neira J, Ceraso DH, Orlandi C, Valentini R, Tiribelli N, Brizuela M, Balasini C, Mare S, Domeniconi G, Ilutovich S, Gómez A, Giuliani J, Barrios C, Valdez P; Registry of the Argentinian Society of Intensive Care SATI. Pandemic 2009 influenza A in Argentina: A study of 337 patients on mechanical ventilation. Am J Respir Crit Care Med. 2010 Jul 1;182(1):41–8. doi: 10.1164/201001-0037OC. Epub 2010 Mar 4. PMID: 20203241.

renal disease, and receipt of mechanical ventilation, vasopressors, renal replacement therapy, or vasodilator therapy, among several others.

8.7 Long COVID and Ultrasound

As the pandemic progresses, chronic health outcomes associated with acute COVID are becoming more frequent and are classified under the "long COVID" category. While the acute phase of COVID-19, which is often mild or moderate, usually lasts one to two weeks, the symptoms of long COVID can last for several months and often appear to be nonspecific and not restricted to organ systems. Clarifying the underlying causes and developing a meaningful, effective, and focused diagnostic algorithm are critical. Lung ultrasound will almost certainly be an essential consideration in the care of these patients, and it has been shown to be useful for the diagnosis of COVID-19 in the acute stage of the disease.[18] Clofent et al. were able to show, in their study of 352 adult long-COVID patients with a long-term follow-up of 2–5 months, that lung ultrasound could be implemented as a first-line diagnostic procedure in the treatment course.[19] It was shown that the outcome

18. Oks M, Cleven KL, Cardenas-Garcia J, Schaub JA, Koenig S, Cohen RI, Mayo PH, Narasimhan M. The effect of point-of-care ultrasonography on imaging studies in the medical ICU: A comparative study. Chest. 2014 Dec;146(6):1574–1577. doi: 10.1378/chest.14-0728. PMID: 25144593.
19. Clofent D, Polverino E, Felipe A, Granados G, Arjona-Peris M, Andreu J, Sánchez-Martínez AL, Varona D, Cabanzo L, Escudero JM, Álvarez A, Loor K, Muñoz X, Culebras M. Lung Ultrasound as a First-Line Test in the Evaluation of Post-COVID-19 Pulmonary

of lung ultrasound in adults could produce good discrimination between patients with persistent abnormalities compared with high-resolution CT. However, in this population, Clofent et al. found a high rate of interstitial lung disease after acute COVID-19 infection. On the other hand, according to a study conducted on children and adolescents by Gräger et al.,[20] lung ultrasound is not suitable as a standard in the follow-up of long-COVID patients without initial pneumonia or pathological findings in an existing baseline examination. They concluded that better standard examination protocols need to be established for this patient group. Nevertheless, lung ultrasound is an important diagnostic tool for the lung because of its radiation-free nature, rapid availability, and increasing establishment in practice (with increasing experience of the examiners). In summary, much is yet to be discovered from a more adequate evaluation and treatment of worsening patients with COVID-19.

8.8 *Self-Assessment*

1. What is referred to as a "sliding lung sign"?
2. What do "A-lines" and "B-lines" represent?

Sequelae. Front Med (Lausanne). 2022 Jan 13;8:815732. doi: 10.3389/fmed.2021.815732. PMID: 35096906; PMCID: PMC8794580.

20. Gräger S, Pfirschke R, Lorenz M, Vilser D, Krämer M, Mentzel HJ, Glutig K. Lung ultrasound in children and adolescents with long-term effects of COVID-19: Initial results. Front Pediatr. 2023 Mar 24;11:1112881. doi: 10.3389/fped.2023.1112881. PMID: 37033176; PMCID: PMC10080098.

3. Why have handheld ultrasound units become popular in primary care in recent years?

8.9 *Further Readings*

1. Lichtenstein DA. Lung ultrasound in the critically ill. An Intensive Care. 2014 Jan 9;4(1):1. doi: 10.1186/2110-5820-4-1. PMID: 24401163; PMCID: PMC3895677.
2. Lichtenstein DA. Current Misconceptions in Lung Ultrasound: A Short Guide for Experts. Chest. 2019 Jul;156(1):21–25. doi: 10.1016/j.chest.2019.02.332. Epub 2019 Mar 11. PMID: 30872018.
3. Lichtenstein D, Mézière G, Biderman P, Gepner A, Barré O. The comet-tail artifact. An ultrasound sign of alveolar-interstitial syndrome. Am J Respir Crit Care Med. 1997 Nov;156(5):1640–6. doi: 10.1164/ajrccm.156.5.96-07096. PMID: 9372688.
4. Duggan NM, Shokoohi H, Liteplo AS, Huang C, Goldsmith AJ. Best Practice Recommendations for Point-of-Care Lung Ultrasound in Patients with Suspected COVID-19. J Emerg Med. 2020 Oct;59(4):515–520. doi: 10.1016/j.jemermed.2020.06.033. Epub 2020 Jun 12. PMID: 32713618; PMCID: PMC7290164.
5. Cascella M, Rajnik M, Aleem A, et al. Features, Evaluation, and Treatment of Coronavirus (COVID-19) [Updated 2023 Aug 18]. In: StatPearls [internet].

Treasure Island (FL): StatPearls Publishing; 2023 Jan–. Available from: https://www.ncbi.nlm.nih.gov/books/NBK554776/

9. Abdominal Ultrasound

9.1 Learning Objectives

After reviewing this chapter, you should be able to do the following:

1. Become familiar with some of the anatomical and physiological features of the various organs in the abdomen.
2. Understand how ultrasound is a valuable tool to assess various features and conditions related to those organs.

9.2 Introduction

As with other areas of the body, abdominal ultrasound is done on different levels for various structures within the abdomen. This chapter will be organized by ultrasound principles. It will provide information on the nuances that different organs and different investigations dictate. Ultrasounds in the abdominal area can be performed to look for abnormal structural growths, gallstones, or problems in other places like the liver, kidneys, pancreas, or spleen. This chapter will primarily focus on renal-, gallbladder-, and liver-related discussions.

9.3 Renal Ultrasound

Kidneys are interesting because understanding renal function and disease involves several disciplines, including chemistry, biology, and physics. In clinical medicine, we often analyze renal function using standard lab tests such as urinalysis, serum creatinine, and serum blood urea nitrogen (BUN). These tests give us clues about both normal function and etiologies of pathology. Examples of how the three tests help us begin abound: too much protein in the urine in the shape of a glomerulus (called "casts") may indicate immune-mediated glomerulonephritis, too-high creatinine alone may indicate diffuse renal failure and intravascular fluid overload, and an altered creatinine-BUN ratio may indicate intravascular fluid depletion.

In a typical ultrasound fashion, we will look at the end anatomic result of renal function and formulate a theory of a patient's condition. Figure 9-1 shows an ultrasonography scan of the longitudinal view of the usual left kidney.

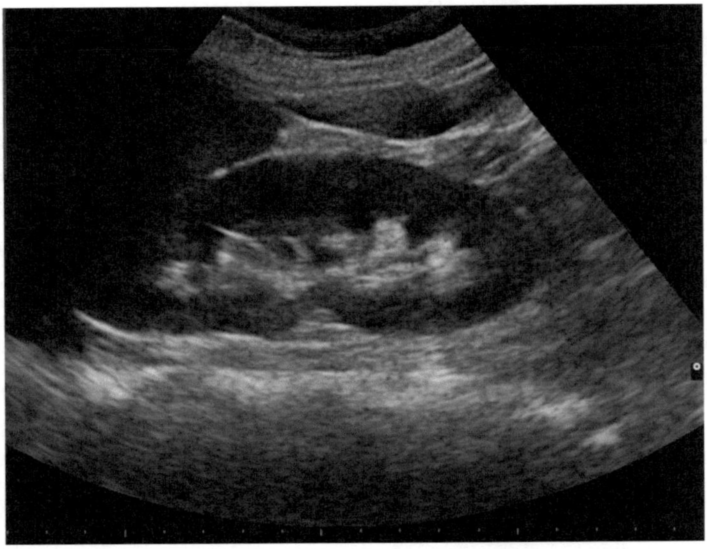

Figure 9-1: Ultrasound image of the longitudinal view of the normal left kidney. Kidney ultrasound 110315132820 1329070 by Nevit Dilmen licensed under CC BY-SA 3.0

Although it might be challenging to view a nephron or its functional parts on ultrasound, we can observe the results of renal malfunction with our gross anatomic view and make some rapid observations that may help a patient. The fascia is the outer fibrous covering of many organs, and Gerota's fascia is the one that surrounds the kidneys and adrenal glands. This fascia is particularly dense and hyperechoic, most often producing a bright reflection (white) back onto the screen in the B-mode. From this distinct outline, we can determine the size, shape, location, and consistency of the surrounding structures near the kidney. The kidney measures approximately 11–14 cm in length, 6 cm in width, and 4 cm in thickness. You may measure these at first, but you may soon only estimate the size visually. Other hyperechoic structures typically surround the kidney. The liver is located superior to the right kidney, and the spleen is located superior to the left kidney. The kidneys are retroperitoneal, or behind the peritoneal cavity. Difficulties in visualizing a kidney are due to the presence of air-filled lungs superior to it as well as the ribs. Air, of course, disperses the ultrasound waves so that reflection is complex. Ribs cause shadowing, which may completely prohibit your initial viewing attempts. Because the liver does not fully extend to the left side, the left kidney is often partially covered by the thoracic cavity and more challenging to visualize.

Other than size and shape, a general clinician ultrasound exam may include lobules, evaluation of cortex thickness, evaluation of the renal pelvis, a survey to evaluate hyperechoic renal calculi (kidney stones), and observation of the vessels entering and leaving the renal pelvis (renal arteries, renal veins, and ureters). If atrophy is noted or a patient has marked hypertension, renal artery blood flow velocity is calculated via Doppler technology to determine if there is renal artery stenosis. This latter exam is usually outside the realm of general clinical ultrasound. It may be best for the general ultrasound clinician to refer this exam to those who do the exam often.

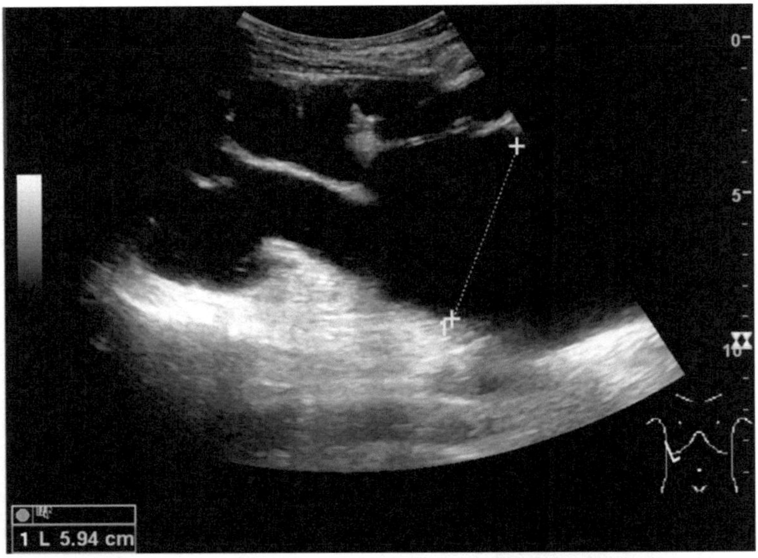

Figure 9-2: Ultrasonography of end-stage hydronephrosis. End-stage hydronephrosis with cortical thinning by Hansen KL, Nielsen MB, Ewertsen C licensed under CC BY 4.0

Large kidneys, or hydronephrosis, may be due to congenital variation, but this most often is a condition due to distal obstruction. Figure 9-2 shows an ultrasound image of end-stage hydronephrosis. Most often, there is only unilateral obstruction of the ureter. As the kidney continues to make urine in the presence of ureteral obstruction, there is backflow pressure and renal swelling. This is often seen in patient care with ureteral calculi obstructing a ureter. Hematuria (blood in the urine), unilateral pain, and unilateral hydronephrosis are diagnostic of ureteral calculi, even if a calculus is too small to be seen (usually less than 3 mm). Treatment is begun based on this clinical presumption.

Choices to diagnose kidney and ureteral stones include intravenous pyelogram (IVP), plain radiographs, CT scans (which allow the synthesis of a 3D picture from multiple radiologic views), and ultrasound. Plain radiography, or shooting an X-ray through the abdomen, is the oldest evaluation method but is still used. Often hydronephrosis and occasionally an actual stone may be visualized. This method is often used to follow a visible stone over several days. IVP production involves injecting dye into a patient's vein and taking serial plain X-rays to observe the flow of dye through the kidney and ureter. This method has lost great popularity due to the dye load on the kidney occasionally causing renal malfunction and less accuracy in diagnosis. CT technology is fast, does not require dye in this particular study, and is most accurate. The clarity of this technology is evident to the most inexperienced patient. Even a very small stone with a typical size of 1 mm may be measured more accurately. The stone may be more easily seen even if it does not contain calcium to reflect ultrasound waves. CT is used most often, but ultrasound is becoming more popular due to cost and the lack of ionizing radiation. Figure 9-3 shows an ultrasound image of a renal stone located at the pyeloureteral junction.

Over 70 million CT scans are performed in the United States every year.[1] The malignant potential of CT scans was most famously brought to the forefront in 2007 by David Brenner and Eric Hall in the *New England Journal of Medicine.*[2] Determining the medical cost of a CT scan is also a complex issue. There is a wide range of costs for CT imaging, typically running from $900 to $3,000. It is conceded by most that clinician-generated ultrasound avoids both of these menacing issues.

1. Brenner DJ. Slowing the increase in the population dose resulting from CT scans. Radiat Res. 2010 Dec;174(6):809–15. doi: 10.1667/RR1859.1. Epub 2010 Aug 23. PMID: 20731591.
2. Brenner DJ, Hall EJ. Computed tomography—an increasing source of radiation exposure. N Engl J Med. 2007 Nov 29;357(22):2277–84. doi: 10.1056/NEJMra072149. PMID: 18046031.

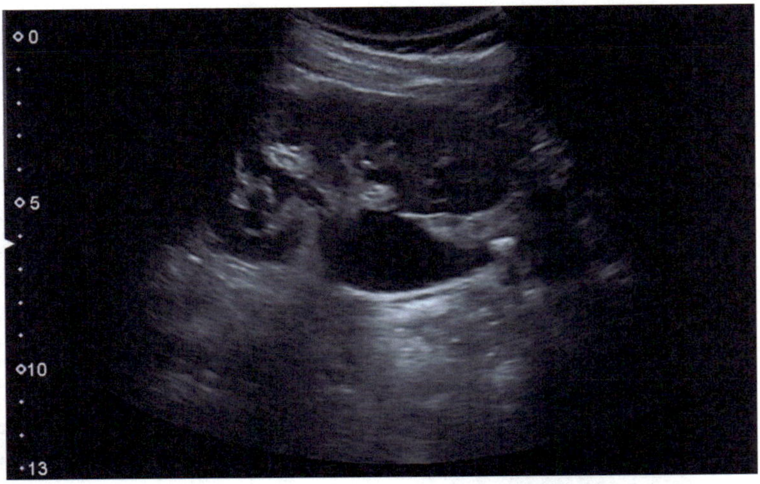

Figure 9-3: Ultrasound scan of renal stone located at the pyeloureteral junction. Renal stone located at the pyeloureteric junciton with accompanying hydronephrosis by Hansen KL, Nielsen MB, Ewertsen C licensed under CC BY 4.0

Clinically, significantly small bilateral kidneys with a thin cortex may indicate chronic renal disease from a diffuse process such as glomerulonephritis or chronic urinary tract infections causing scarring. This condition is distinct from a single small kidney and indicates a localized problem, such as decreased blood flow to only one kidney, known as *renal artery stenosis.*

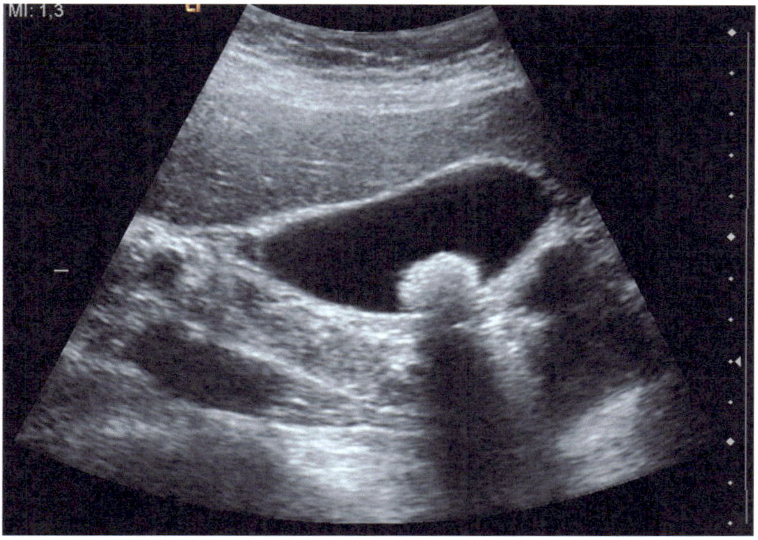

Figure 9-4: Ultrasound image of gallbladder stone. Ultrasound image of gallbladder stone Gallstone 091937515 by Nevit Dilmen licensed under CC BY-SA 3.0

9.4 Gallbladder Ultrasound

Gallbladder ultrasounds are standard diagnostic investigations that can be done by primary care and emergency providers. The capsule of the gallbladder, with its fluid-filled contents, often makes a very clearly defined reflective surface. Figure 9-4 shows an ultrasound image of a gallbladder stone. Calculi, often called "gallstones," may be seen within the gallbladder because of the high density and reflectivity of the discrete objects sitting in often clear fluid. Another feature called "shadowing" is helpful in diagnosis. *Shadowing* refers to the sharply demarcated darkness that is under the gallstone. Not all gallstones cause disease or need to be addressed. Other features that can be noted on ultrasound and can indicate pathology or a diseased state in a patient include the location of the calculi, the size of the gallbladder, inflammation of the gallbladder, and acute cholecystitis.

9.4.1 Location of the Calculi

Gallstones close to the gallbladder neck are speculated to have more potential to eventually migrate through the gallbladder neck, into the cystic duct, and even into the common bile duct. Gallstones are often more challenging to visualize in the cystic duct or common bile duct due to their small size and coexisting bowel gas reflecting the ultrasound beam before it can reach the gallbladder. Other modalities are often needed to assist in diagnosis.

9.4.2 The Size of the Gallbladder

The gallbladder contracts with cholecystokinin hormone, stimulating the vagus nerve in the parasympathetic nervous system. A small gallbladder may indicate chronic inflammation and scarring. A large gallbladder may also be pathological, even indicating rare gallbladder cancer. The gallbladder size is recorded by manipulating the measurement function on the machine.

9.4.3 Inflammation of the Gallbladder

The wall of the gallbladder can be thickened from inflammation. Greater than 0.3 cm wall thickness is often defined as being pathologically thickened.

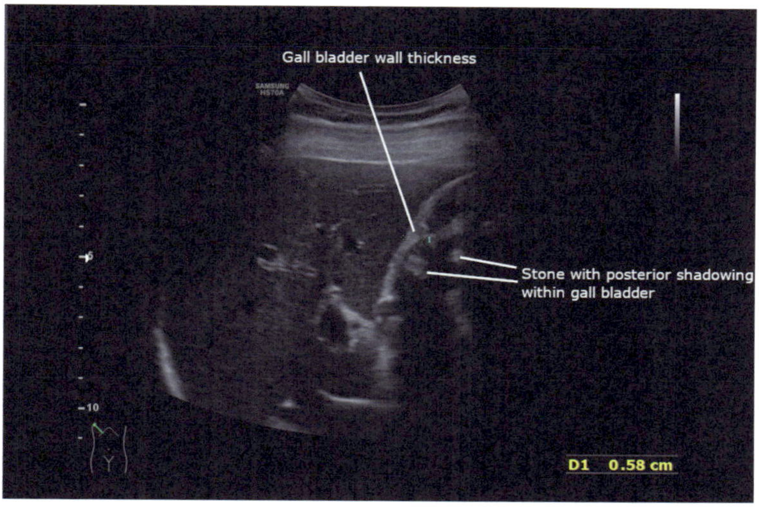

Figure 9-5: Acute cholecystitis as seen on the ultrasound axial view. Acute cholecystitis as seen on ultrasound axial view by Cerebisae licensed under CC BY-SA 4.0

9.4.4 Acute Cholecystitis

Surrounding a thickened wall can also be inflammatory fluid, indicating an acute inflammatory response or "acute cholecystitis." This inflammatory fluid will be seen as a dark area outside the (usually thickened) gallbladder wall, as shown in Figure 9-5.

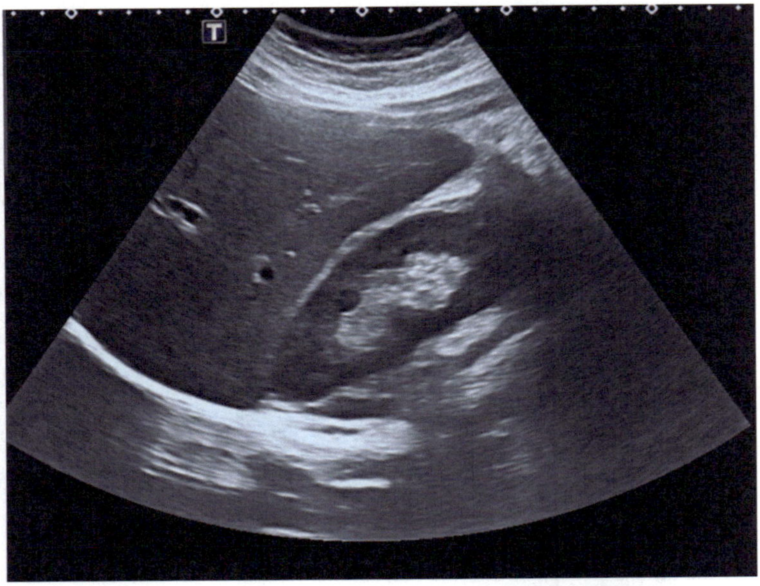

Figure 9-6: Abdominal ultrasound showing the right lobe of the liver and right kidney. Ultrasound liver right lobe and right kidney by Ptrump16 licensed under CC BY-SA 4.0

9.5 Liver Ultrasound

Figure 9-6 is an abdominal ultrasound showing the right lobe of the liver and right kidney. Compared to the gallbladder, kidney, intestine, or bladder, the liver's thin covering makes it less distinctive. Distinctive features include the following:

- Vascular analysis—Cirrhosis is a pathologic liver condition arising from inflammation and subsequent scarring. This scarring impacts the ease with which blood from the digestive tract and other places flows into the liver. The portal venous backflow causes portal dilation and portal hypertension. This is evaluated by measuring the portal venous velocity using the

Doppler and color Doppler functions. There are ultrasound-guided procedures that are used to relieve vascular congestion.

- The duct system in the liver (intrahepatic duct)—Because of the higher lipid contact (and therefore the higher sonar beam reflectivity) of the biliary duct walls, they can be more easily visualized, making diameter determination of the ductal system more convenient. A dilated intrahepatic duct system can arise from a ductal obstruction. Frequently, obstructions may also be calculi from the gallbladder or cancers of organs farther "downstream" in the hepatic system, such as cancer of the pancreas.
- The fluids and reflective surface—The liver is a reflective surface for ultrasound beams. Pathologic fluid in the abdomen may be chronic, such as the fluid from chronic liver disease (known as *ascites*). It may also be acute, such as an acute intra-abdominal bleed from a motor vehicle accident. The capsular surfaces between the liver and the kidney are usually touching in a healthy individual. As discussed in the earlier chapters, the pressure of abnormal fluid forces the hepatic and renal surfaces apart, creating a new dark space referred to as Morrison's pouch. In a patient with unlikely chronic ascites due to acute abdominal trauma, an emergency surgery may be performed to stop active bleeding into the abdomen. The emergency ultrasound exam of a patient who has received abdominal trauma is referred to as eFAST, indicating an **E**xtended **F**ocused **A**ssessment with **S**onography in **T**rauma. The original basis of this exam is the attempt to find abnormal (dark) fluid between two highly reflective surfaces that are generally close or even touching. Other than Morrison's pouch, other areas with potential spaces are between the pericardial and myocardial heart surface, the left kidney and the spleen, and the bladder and uterus in females. In females who have had a hysterectomy and in typical male anatomy, the area under the dense capsule of the urinary bladder can show a dark region of collected fluid. FAST and eFAST exams will be

discussed in a little more detail in Chapter 11.

- Abnormal liver size—The liver can be pathologically large from several diseases, including acute hepatitis and liver cancer. The size of the liver is not usually the first clue to either diagnosis but a noted pathology when the disease is already suspected.

9.6 *Self-Assessment*

1. What causes difficulties in visualizing the kidneys in an ultrasound scan?
2. Other than size and shape measurements, what other ultrasound evaluations may a general clinician perform?
3. What does "shadowing" refer to in the context of an ultrasound scan of the gallbladder?
4. How are Doppler and color Doppler functions useful in vascular analysis of the liver?

9.7 *Further Readings*

1. Hagopian EJ, Machi J. Abdominal Ultrasound for Surgeons. [place unknown]: Springer; 2014. 242 p.
2. Brenner DJ. Minimising medically unwarranted computed tomography scans. Ann ICRP. 2012 Oct–Dec;41(3–4):161–9. doi: 10.1016/j.icrp.2012.06.004.

Epub 2012 Aug 22. PMID: 23089015.
3. Caraiani C, Yi D, Petresc B, Dietrich C. Indications for abdominal imaging: When and what to choose? J Ultrason. 2020;20(80):e43–e54. doi: 10.15557/JoU.2020.0008. Epub 2020 Mar 31. PMID: 32320166; PMCID: PMC7266076.
4. American Institute of Ultrasound in Medicine. AIUM practice guideline for the performance of pelvic ultrasound examinations. J Ultrasound Med. 2010 Jan;29(1):166–72. doi: 10.7863/jum.2010.29.1.166. PMID: 20040793.

10. Vascular Sonography

10.1 Learning Objectives

After reviewing this chapter, you should be able to do the following:

1. View and identify the anatomical structures of the venous systems and the corresponding ultrasound images.
2. Explore the arterial system and the corresponding ultrasound images.

10.2 Introduction

The topics covered in this section will include some of the ultrasound basics of the venous and arterial systems, including transcranial, carotid, aorta, and lower-extremity ultrasound imaging.

10.3 The Venous System

The primary physiologic functions of the venous system are to return the deoxygenated blood to the heart, thermoregulate, store blood (at any instance, the venous system contains up to 70% of the circulating blood), and regulate the cardiac output. It is divided into three systems: superficial, perforating, and deep veins. Figure 10-1 shows the anatomy of the venous system. Blood flows from the superficial to deep veins through branching perforating veins. The deep veins usually follow the arteries in the same areas and often have similar names. For example, the femoral vein is beside the femoral artery. The deep venous system eventually returns blood to the right side of the heart. Since the venous system is usually a low-pressure system, veins have bicuspid valves to allow flow in one direction from superficial to deep (the foot is the exception) and from distal to proximal. Muscular contraction helps with venous flow, such as in the calf muscle pump in the leg.[1]

1. Meissner MH, Moneta G, Burnand K, Gloviczki P, Lohr JM, Lurie F, Mattos MA, McLafferty RB, Mozes G, Rutherford RB, Padberg F, Sumner DS. The hemodynamics and diagnosis of venous disease. J Vasc Surg. 2007 Dec;46 Suppl S:4S-24S. doi: 10.1016/j.jvs.2007.09.043. PMID: 18068561.

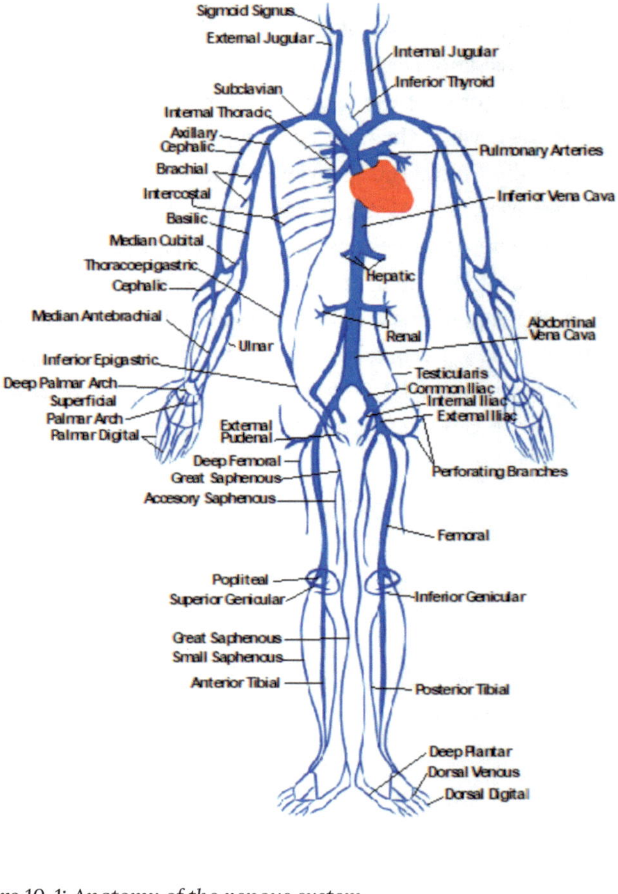

Figure 10-1: Anatomy of the venous system.

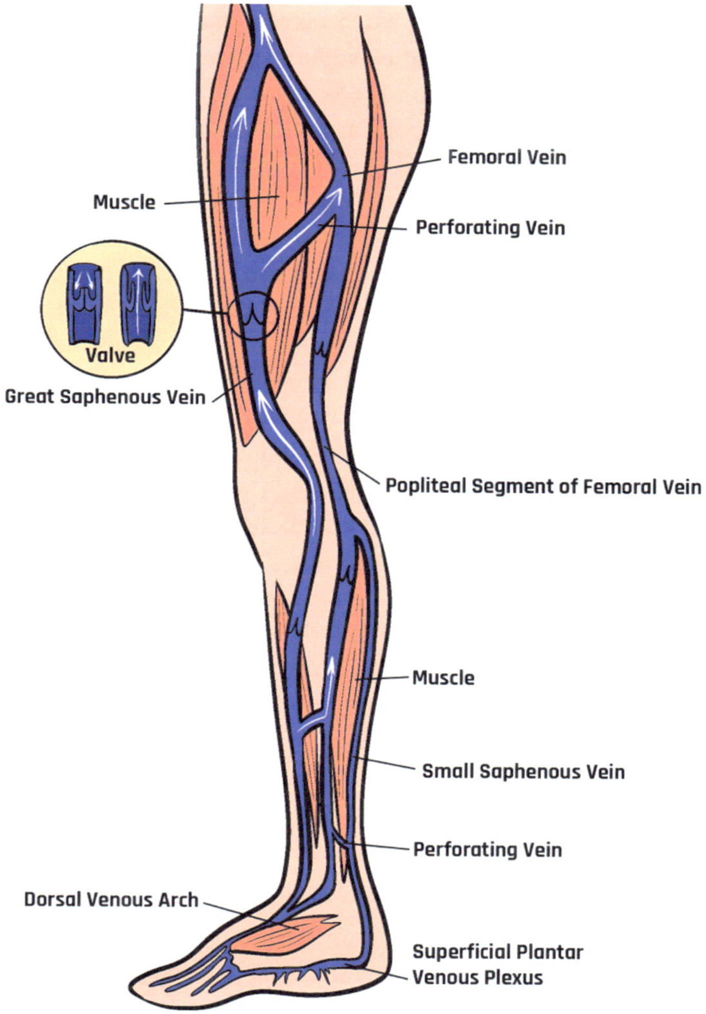

Figure 10-2: Valves in the venous system of the leg.

Venous pathophysiology has many etiologies, such as trauma and genetic predisposition, and can occur when outflow is impaired by dysfunctional valves, resulting in retrograde flow and causing a condition known as chronic venous insufficiency. Vein thrombosis is another condition with many hereditary and acquired etiologies, such as trauma or prolonged immobilization. Deep vein thrombosis is especially important to evaluate and treat.[2]

The great saphenous vein (GSV) is the longest vein in the human body, as shown in Figure 10-2. It originates in the medial aspect of the foot as part of the dorsal arch. It continues proximally along the medial aspect of the foot and passes anterior to the medial malleolus on the tibia. It ascends along the medial aspect of the leg between the superficial and deep fascia. It typically has 10–20 valves and terminates at the saphenofemoral junction (SFJ). Once flow enters the femoral vein, it is in the deep venous system. Venous anatomy can vary from individual to individual. However, the GSV typically has branching superficial veins, such as the anterior and posterior accessory saphenous veins in the thigh.[3]

The small saphenous vein (SSV) is the second most significant superficial vein that joins the dorsal venous arch in the lateral aspect of the foot. It ascends proximally behind the lateral malleolus and terminates into the deep popliteal vein, although this is highly variable and can extend into the thigh. The SSV typically has 9–12 valves. Like the GSV, the SSV lies between the superficial and deep fascia and can have many branching superficial veins. Perforating veins connect superficial to deep veins. They usually contain a bicuspid valve.[4]

2. Schellong S, Schwarz T. Peripheral Venous Anatomy and Physiology. In: Lanzer P, Topol EJ, editors. Pan Vascular Medicine. Berlin, Heidelberg: Springer; 2002. p. 1489–1491. Available from: https://doi.org/10.1007/978-3-642-56225-9_92
3. Schellong S, Schwarz T. Peripheral Venous Anatomy and Physiology. In: Lanzer P, Topol EJ, editors. Pan Vascular Medicine. Berlin, Heidelberg: Springer; 2002. p. 1489–1491. Available from: https://doi.org/10.1007/978-3-642-56225-9_92
4. Schellong S, Schwarz T. Peripheral Venous Anatomy and

The deep venous system includes the common femoral vein, profunda femoral vein, deep femoral vein, popliteal vein, gastrocnemius veins, soleus veins, anterior tibial veins, posterior tibial veins, and peroneal veins. The direction of venous flow is described as antegrade, retrograde, or absent. In both the deep and superficial venous systems, it is essential to check for the following characteristics: compressibility, spontaneous flow, respiratory variation, augmentation, intraluminal defects, and venous reflux.

Compressibility evaluates if the vein collapses by applying downward pressure with the transducer. Typically, it should compress, since it is a low-pressure vessel. A thrombus can occlude the lumen and prevent compression. Spontaneous flow is observed when the blood flow moves actively without external influences, such as an augmentation maneuver. Respiratory variation, also known as phasicity, refers to regular venous flow changes that occur secondary to intrathoracic pressure during breathing cycles. Augmentation is a maneuver that is used to evaluate possible abnormal flow patterns. For example, by squeezing a distal portion in the calf, an increase in venous flow should be observed just proximal to this area. Absent or diminished flow could suggest obstruction, such as in a thrombus formation, and reversal of flow could indicate incompetent venous valves, such as in venous reflux disease.

Intraluminal defects usually describe a thrombus formation within the lumen of the vein. It is crucial to describe the details of the thrombus formation and whether it is obstructive.

Finally, venous reflux describes blood flow going in the wrong direction, usually from incompetent valves. Maneuvers are usually done to augment blood flow to test for reflux, which is significant if it exceeds 0.5 seconds in the superficial venous system,

Physiology. In: Lanzer P, Topol EJ, editors. Pan Vascular Medicine. Berlin, Heidelberg: Springer; 2002. p. 1489–1491. Available from: https://doi.org/10.1007/978-3-642-56225-9_92

0.35 seconds in perforators, and 1 second in the deep venous system.

A complete venous duplex ultrasound of the lower extremities starts with a proximal to distal evaluation of the deep venous system in a transverse (TRV) side-by-side image without compression and with compression (COMP). This is followed by a sagittal (SAG) view with augmentation (AUG) using the color Doppler. The veins that are evaluated in succession include the common femoral vein (CFV), profunda femoral vein (PROF V), femoral vein (FV), popliteal vein (POP V), gastrocnemius vein (GASTROC V), posterior tibial vein (PTV), peroneal vein (PERO V), anterior tibial vein (ATV), great saphenous vein (GSV), and small saphenous vein (SSV). The abbreviations given in parentheses in the last few sentences have been labeled in some of the following ultrasound images for venous system discussion. Figure 10-3 shows a side-by-side transverse ultrasound view of the right common femoral vein without compression and with compression, while Figure 10-4 represents the sagittal view of the right common femoral vein with augmentation.

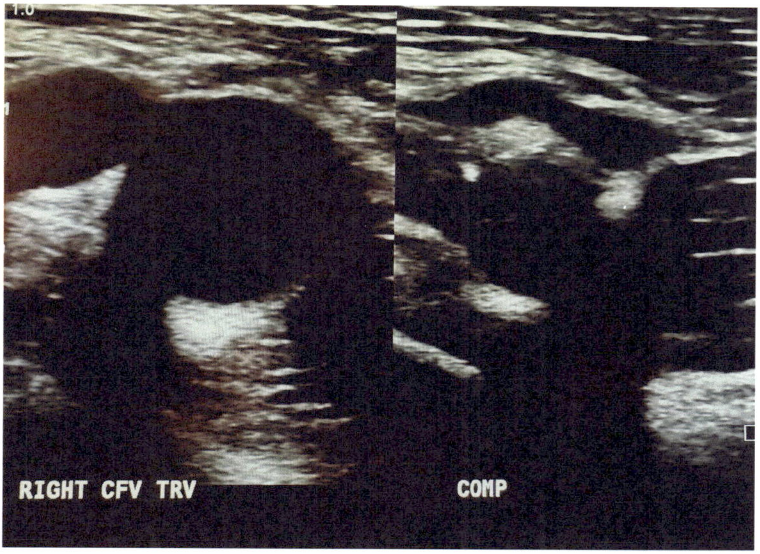

Figure 10-3: Side-by-side transverse ultrasound views of the right common femoral vein without compression and with compression.

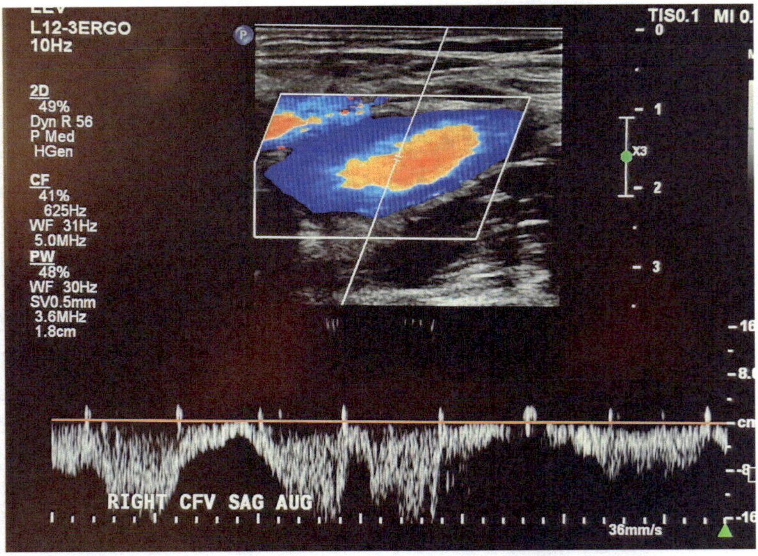

Figure 10-4: Right common femoral vein sagittal view with augmentation.

Figure 10-5 shows a side-by-side transverse ultrasound view of the right profunda femoral vein without compression and with compression, while Figure 10-6 represents the sagittal view of the right profunda femoral vein with augmentation.

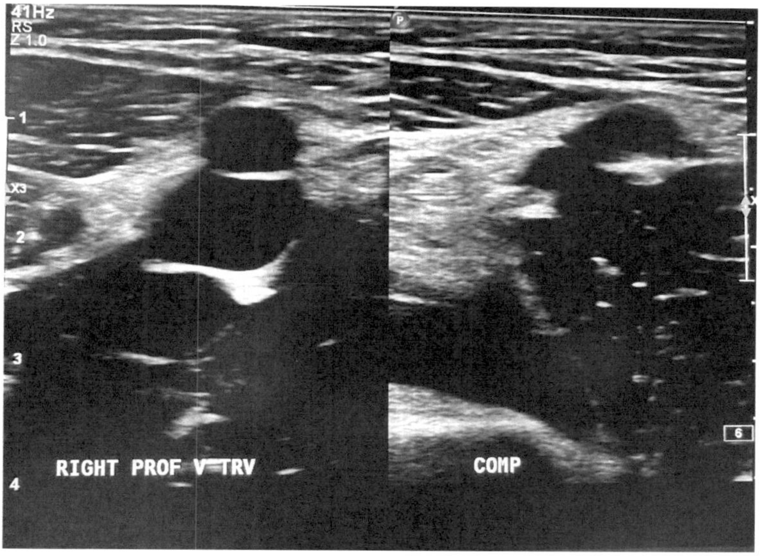

Figure 10-5: Side-by-side transverse image of the right profunda femoral vein without and with compression.

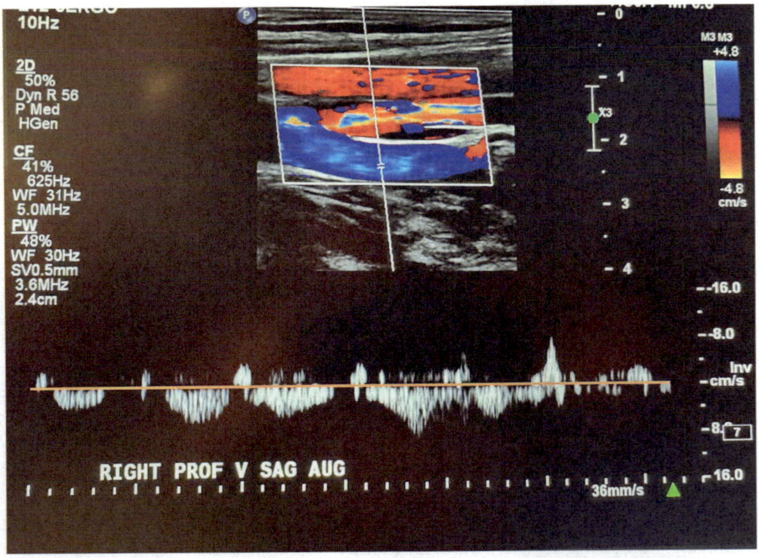

Figure 10-6: Right profunda femoral vein sagittal view with augmentation.

Figure 10-7 shows a side-by-side transverse ultrasound view of the right femoral vein without compression and with compression, while Figure 10-8 represents the sagittal view of the right femoral vein with augmentation.

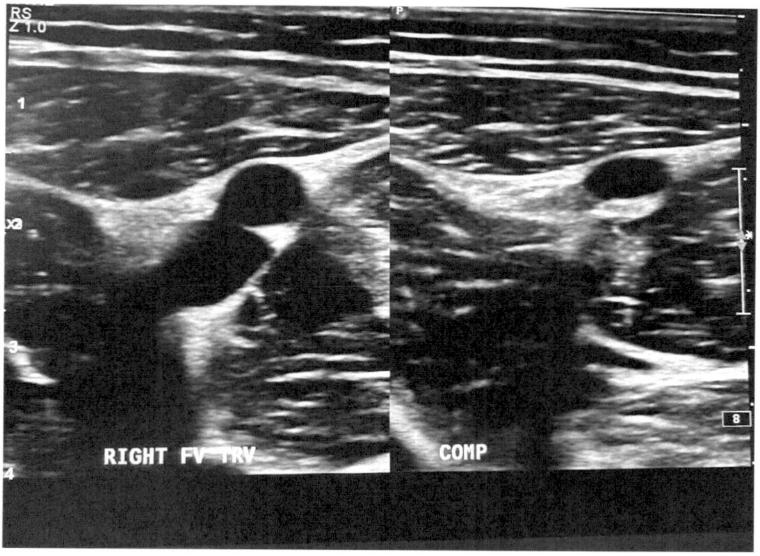

Figure 10-7: Side-by-side right femoral vein transverse view without and with compression.

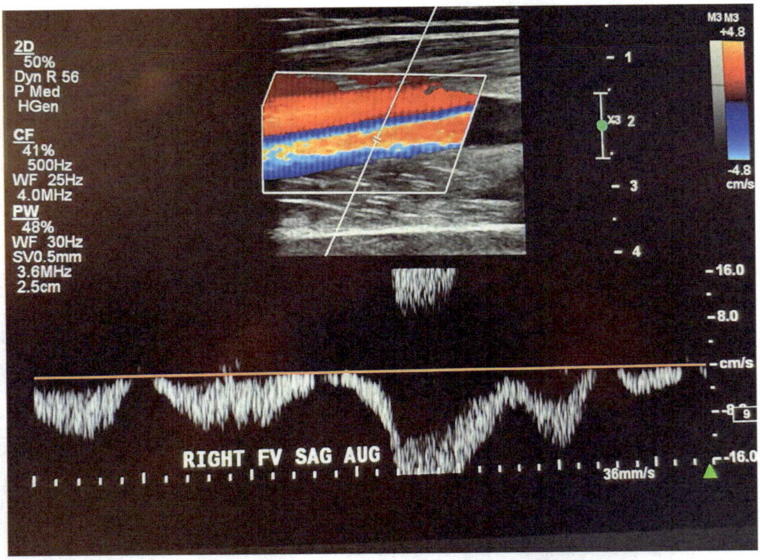

Figure 10-8: Right femoral vein sagittal view with augmentation.

Figure 10-9 shows a side-by-side transverse ultrasound view of the right popliteal vein without compression and with compression, while Figure 10-10 represents the sagittal view of the right popliteal vein with augmentation.

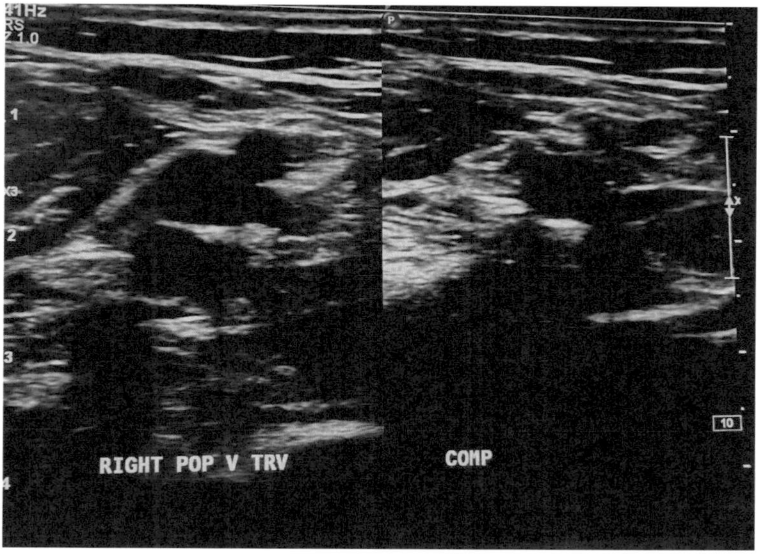

Figure 10-9: Side-by-side right popliteal vein transverse view without and with compression.

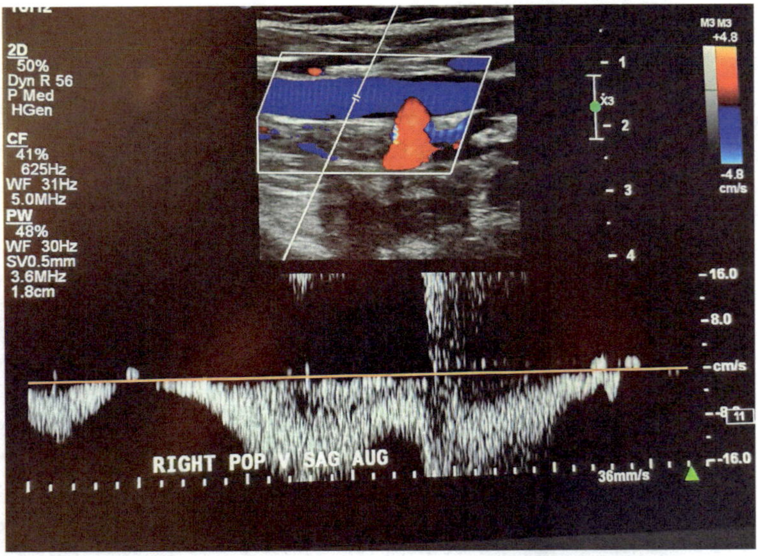

Figure 10-10: Right popliteal vein sagittal view with augmentation.

Figure 10-11 shows a side-by-side transverse ultrasound view of the right gastrocnemius vein without compression and with compression, while Figure 10-12 represents the sagittal view of the right gastrocnemius vein with augmentation.

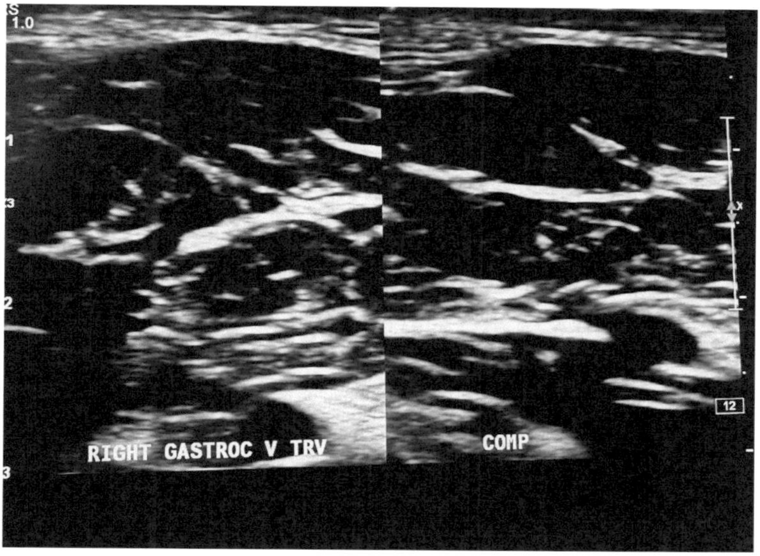

Figure 10-11: Side-by-side right gastrocnemius vein transverse view without and with compression.

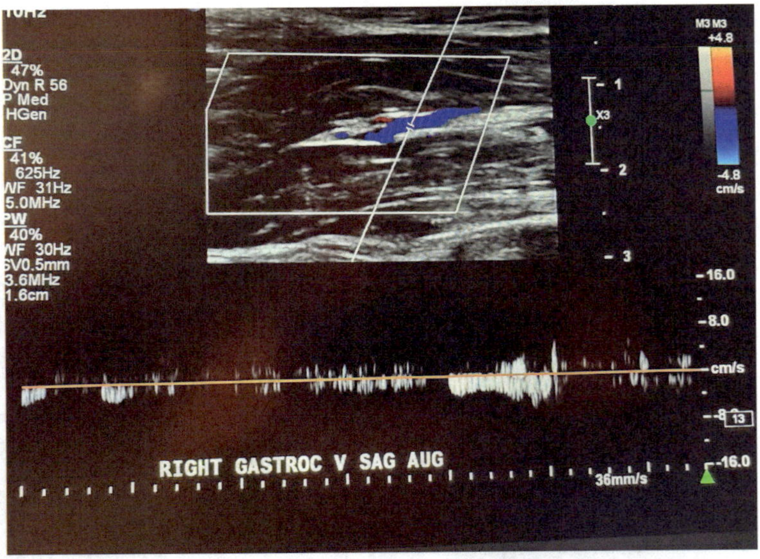

Figure 10-12: Right gastrocnemius vein sagittal view with augmentation.

Figure 10-13 shows a side-by-side transverse ultrasound view of the right posterior tibial vein without compression and with compression, while Figure 10-14 represents the sagittal view of the right posterior tibial vein with augmentation.

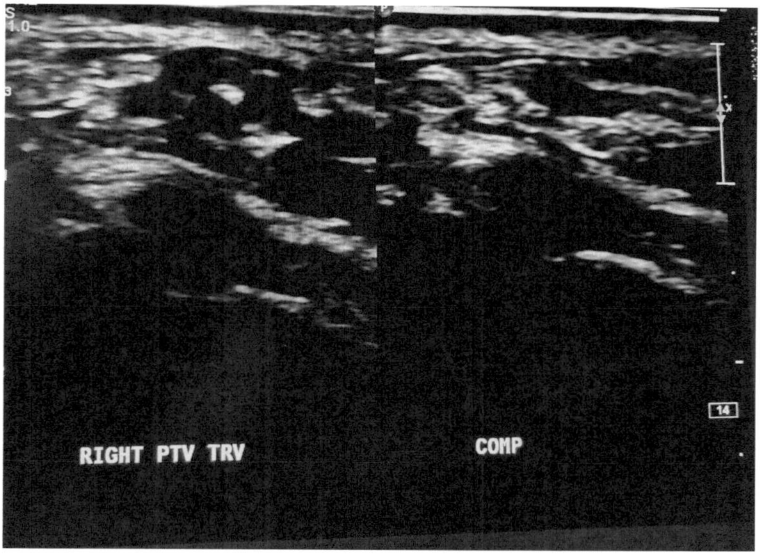

Figure 10-13: Side-by-side right posterior tibial veins transverse view without and with compression.

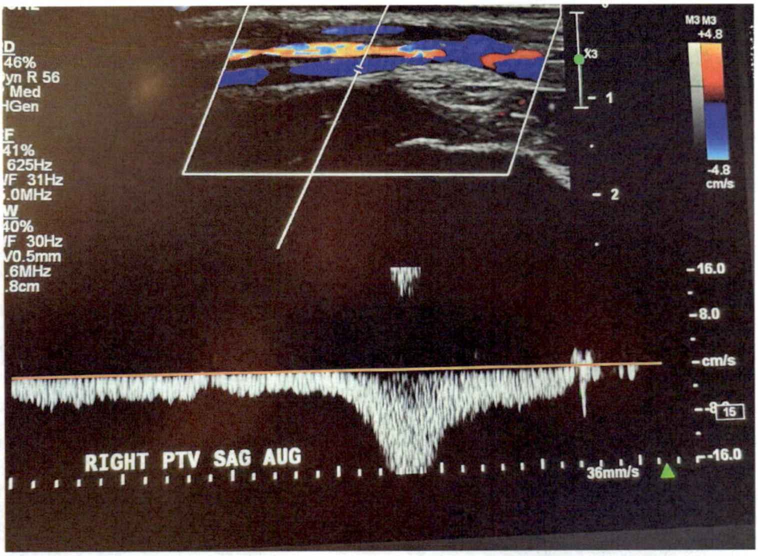

Figure 10-14: Right posterior tibial veins sagittal view with augmentation.

Figure 10-15 shows a side-by-side transverse ultrasound view of the right peroneal vein without compression and with compression, while Figure 10-16 represents the sagittal view of the right peroneal vein with augmentation.

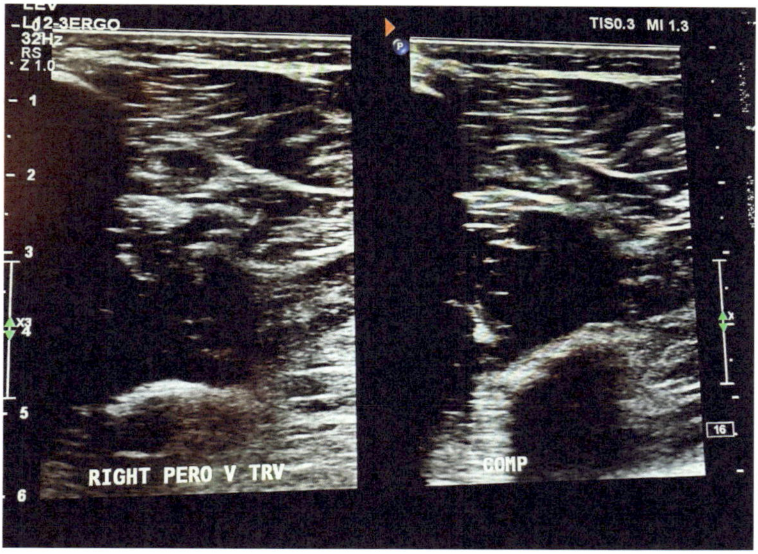

Figure 10-15: Side-by-side right peroneal veins transverse view without and with compression.

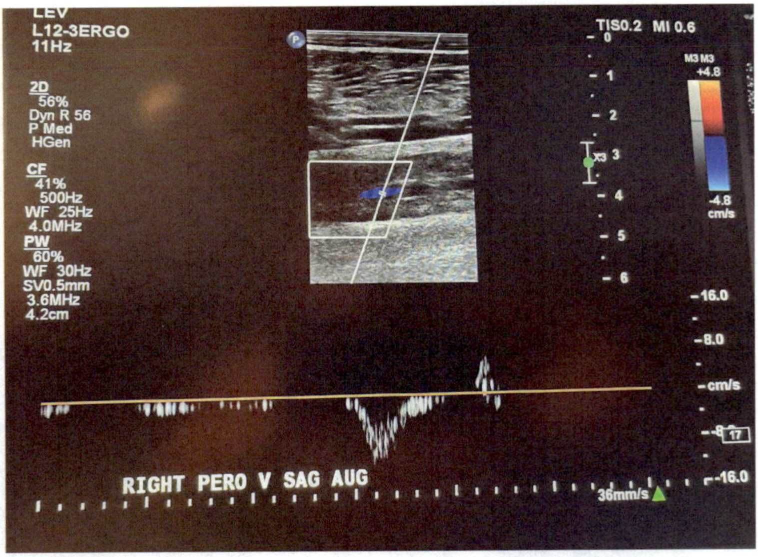

Figure 10-16: Right peroneal vein sagittal view with augmentation.

Figure 10-17 shows a side-by-side transverse ultrasound view of the anterior tibial vein without compression and with compression, while Figure 10-18 represents the sagittal view of the right anterior tibial vein with augmentation.

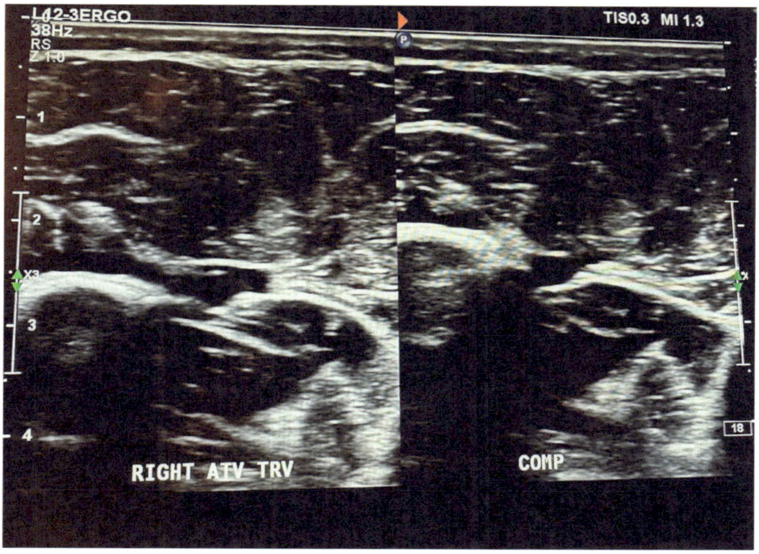

Figure 10-17: Side-by-side right anterior tibial vein transverse view without and with compression.

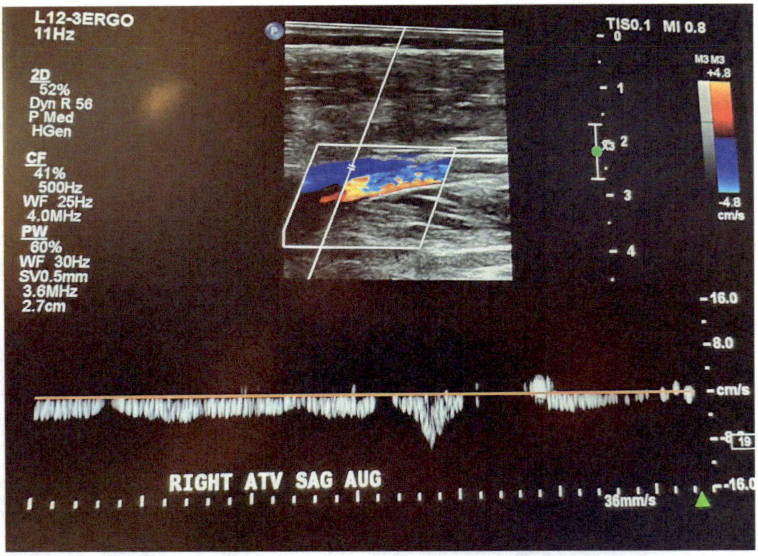

Figure 10-18: Right anterior tibial vein sagittal view with augmentation.

Next, in the complete venous duplex ultrasound, we look at the superficial venous system from proximal to distal, starting with the GSV at the SFJ in the transverse plane with a side-by-side image without and with color Doppler, followed by a sagittal image with augmentation with color Doppler. Figure 10-19 shows a side-by-side transverse ultrasound view of the right GSV at the SFJ, while Figure 10-20 represents the sagittal view of the right GSV at the SFJ with augmentation.

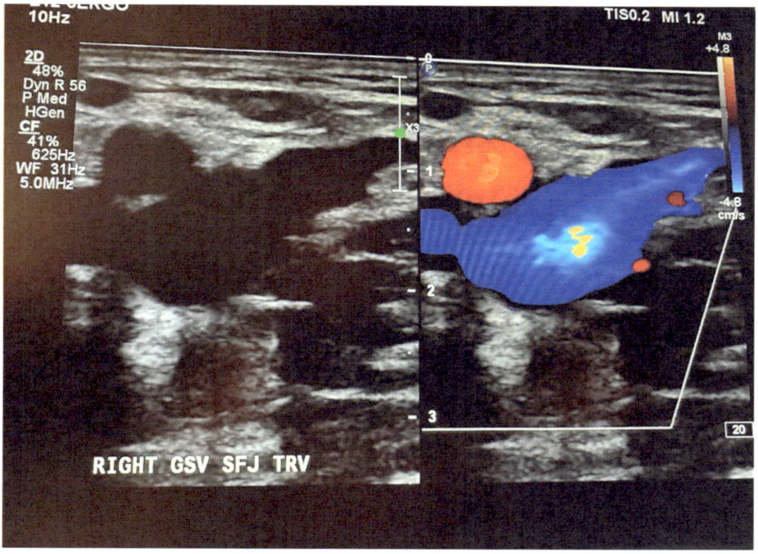

Figure 10-19: Side-by-side right great saphenous vein at the saphenofemoral junction transverse view without and with color Doppler.

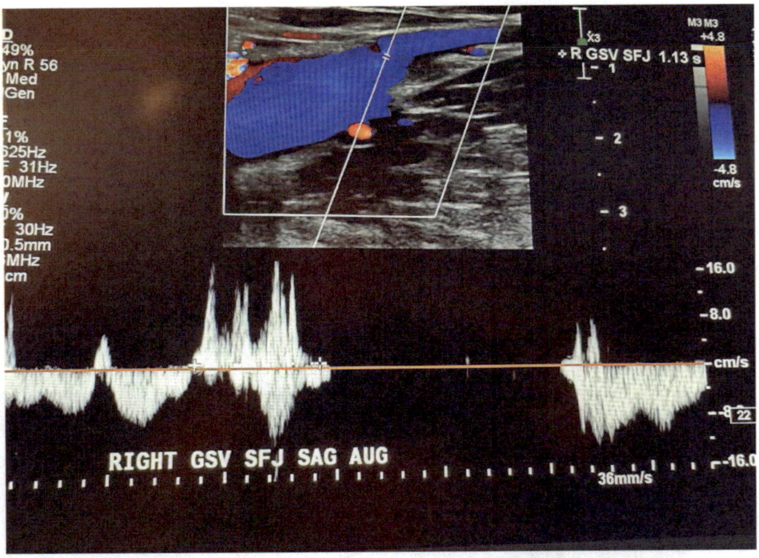

Figure 10-20: Right great saphenous vein at the saphenofemoral junction sagittal view with augmentation.

Next, we continue to follow and evaluate the GSV distally from above the knee (AK) to below the knee (BK) in the transverse plane without color, followed by the sagittal plane with augmentation with color Doppler. The abbreviations AK and BK have been used in the ultrasound images discussed here. Figure 10-21 shows a side-by-side transverse ultrasound view of the right GSV above the knee. Figure 10-22 represents the sagittal view of the right GSV above the knee with augmentation.

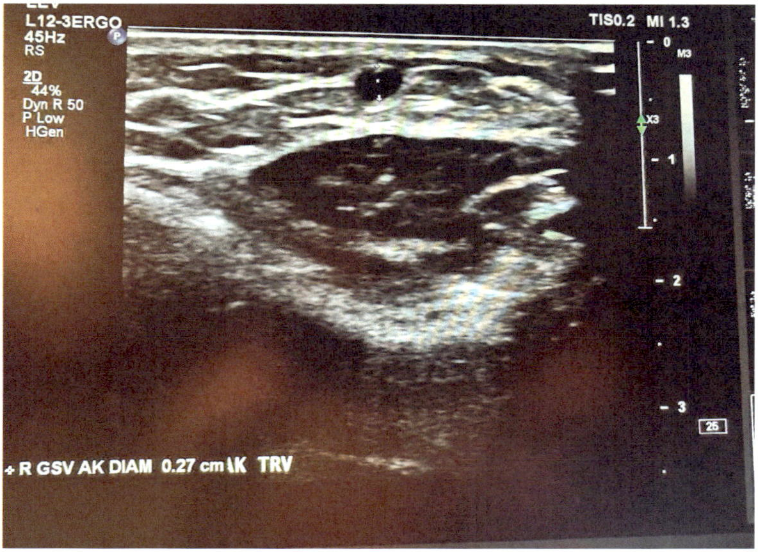

Figure 10-21: Right great saphenous vein transverse view above the knee.

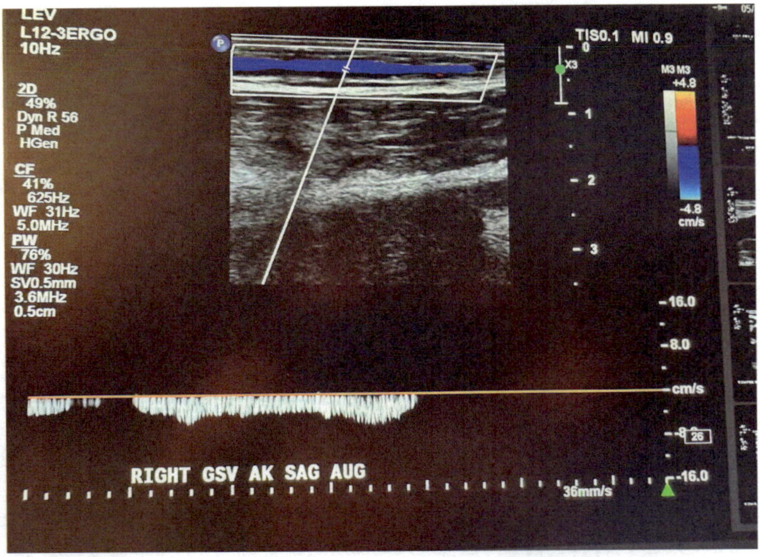

Figure 10-22: Right great saphenous vein above the knee sagittal view with augmentation.

Figure 10-23 shows a transverse ultrasound view of the right GSV below the knee, while Figure 10-24 represents the sagittal view of the right GSV below the knee with augmentation.

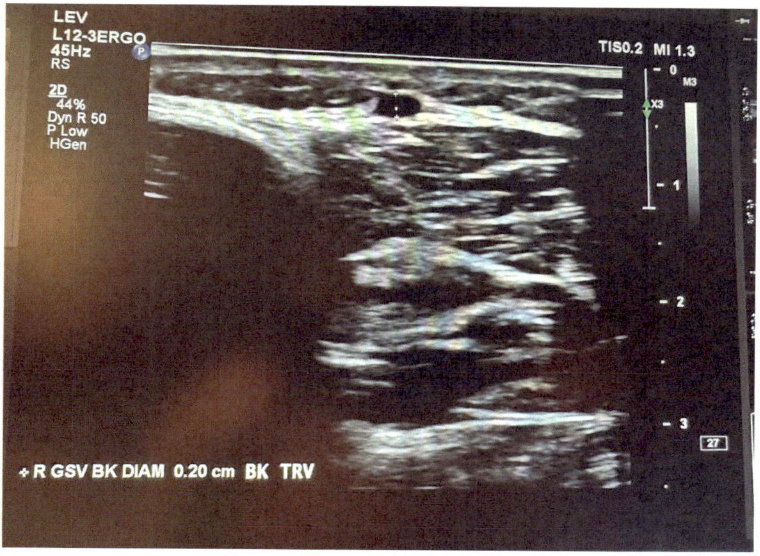

Figure 10-23: Right great saphenous vein below the knee transverse view.

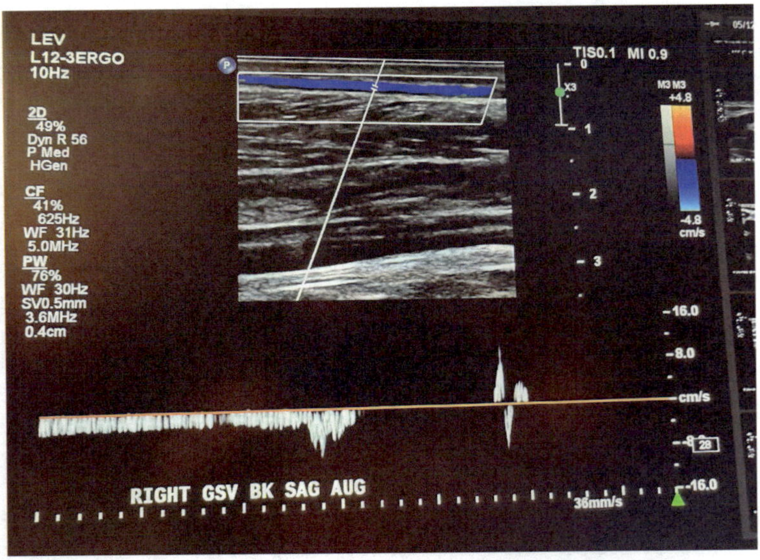

Figure 10-24: Right great saphenous vein below the knee sagittal view with augmentation.

The next superficial vein to be evaluated is the SSV, starting in the popliteal area of the lower extremity and using the same approach as the great saphenous vein. We first start with a transverse image at the saphenopopliteal junction (SPJ) without color Doppler, followed by a sagittal image with augmentation with color Doppler. Figure 10-25 shows a side-by-side transverse ultrasound view of the right SSV at the SPJ, while Figure 10-26 represents the sagittal view of the right SSV at the SPJ with augmentation.

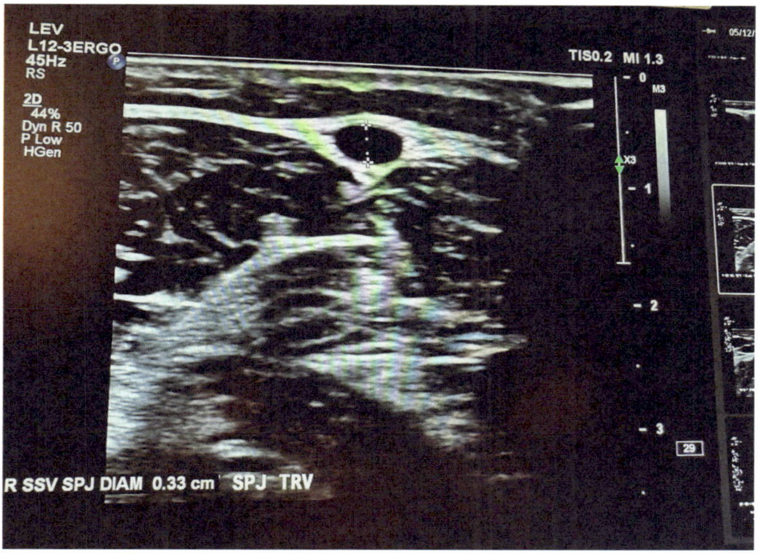

Figure 10-25: Right small saphenous vein at the saphenopopliteal junction transverse view.

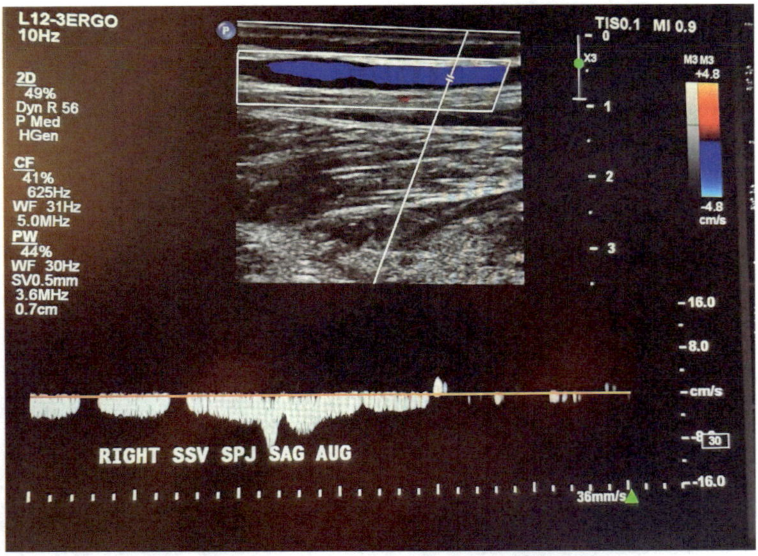

Figure 10-26: Right small saphenous vein at the saphenopopliteal junction sagittal view with augmentation.

Also, anterior and posterior accessory saphenous veins are often evaluated as part of the superficial venous system, and perforating veins that connect superficial to deep veins are often evaluated during the study.

The template shown in the next couple of pages can be used to perform a complete venous duplex Doppler ultrasound examination of the lower extremities.

Date ___________________

U/S technician ____________________

Physician _________________

Complaint______________________________

Comparative study ____________________

	Present	Absent	Right	Left
Leg pain and/or swelling				
Varicose veins				
Leg ulcers				
Deep vein thrombosis				
Pulmonary embolus				
Vein stripping				
Ablation				
Sclerotherapy				
Phlebectomy				
Congestive heart failure				

DEEP VENOUS SYSTEM

Right	Normal compressib-ility	Spontane-ous flow	Respiratory variation	Augmenta-tion	Intraluminal filling defect	Reflux
Common femoral vein						
Profunda femoral vein						
Femoral vein						
Popliteal vein						
Gastrocnemius /soleus vein(s)						
Anterior tibial veins						
Posterior tibial veins						
Peroneal veins						
Left	Normal compressib-ility	Spontane-ous flow	Respiratory variation	Augmenta-tion	Intraluminal filling defect	Reflux
Common femoral vein						
Profunda femoral vein						
Femoral vein						
Popliteal vein						
Gastrocnemius /soleus vein(s)						
Anterior tibial veins						
Posterior tibial veins						
Peroneal veins						

SUPERFICIAL VENOUS SYSTEM

Right	Diameter (cm)	Spontaneous flow	Intraluminal filling defect	Reflux
Accessory saphenous				
Left	Diameter (cm)	Spontaneous flow	Intraluminal filling defect	Reflux
Accessory saphenous				
Right	Diameter (cm)	Spontaneous flow	Intraluminal filling defect	Reflux
GSV—SFJ				
GSV—upper thigh				
GSV—lower thigh				
GSV—below knee				
SSV				
Left	Diameter (cm)	Spontaneous flow	Intraluminal filling defect	Reflux
GSV—SFJ				
GSV—upper thigh				
GSV—lower thigh				
GSV—below knee				
SSV				

Describe significant perforating veins, including size, location, and reflux:

10.4 The Arterial System

Figure 10-27 shows the anatomy of the arterial system, which will be helpful in discussing and understanding various ultrasonography images of the arteries.

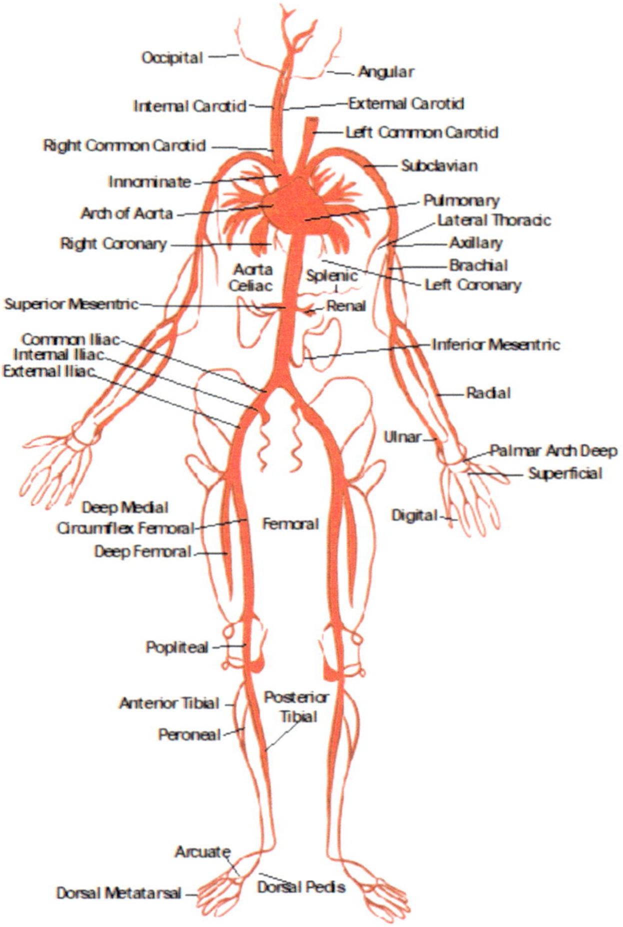

Figure 10-27: Anatomy of the arterial system.

10.4.1 Transcranial Doppler

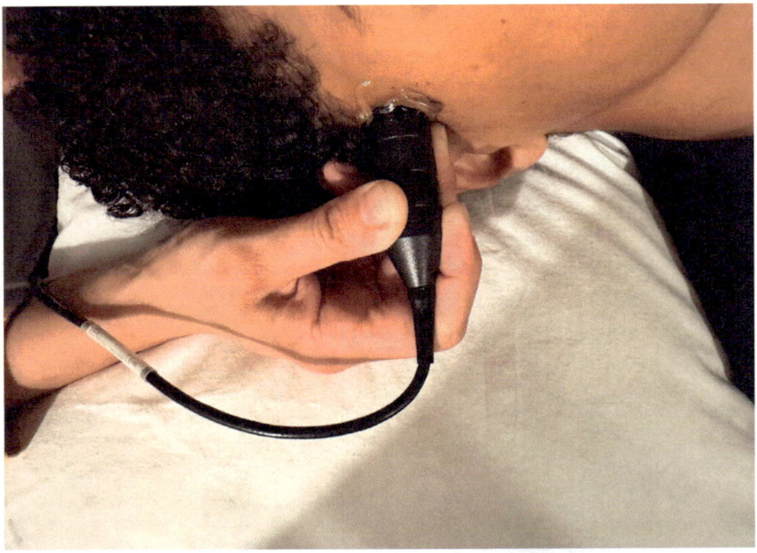

Figure 10-28: Transcranial Doppler probe.

Norwegian physicist Rune Aaslid developed intracranial ultrasound in 1982. Transcranial imaging was subsequently developed by a German neurologist, Ulrich Bogdahn, in 1990. It was the first noninvasive way to evaluate the circle of Willis using a low-frequency transducer (2 MHz).[5],[6]

Transcranial Doppler (TCD) can detect intracranial

5. Davis D. Introduction to Transcranial Doppler Ultrasound [DVD]. St Petersburg (FL): Gulfcoast Ultrasound Institute; 2013 Feb 28–Mar 21.
6. Katz ML, Alexandrov AV. A Practical Guide to Transcranial Doppler Examinations. [place unknown]: Summer publishing; 2003. p. 158.

stenosis, vasospasm secondary to subarachnoid hemorrhage, and arteriovenous malformations and assess suspected brain death. A TCD system usually has a 2 MHz pulsed Doppler with a spectrum analyzer. A typical TCD probe is shown in Figure 10-28. Blood flow in TCD is usually measured in cm/sec, and Figure 10-29 represents a TCD velocity distribution. When evaluating intracranial vessels, it is vital to know the acoustic window, depth, direction of blood flow, velocity, and angle of insonation. One important principle when evaluating pathology is the pulsatility index (PI).

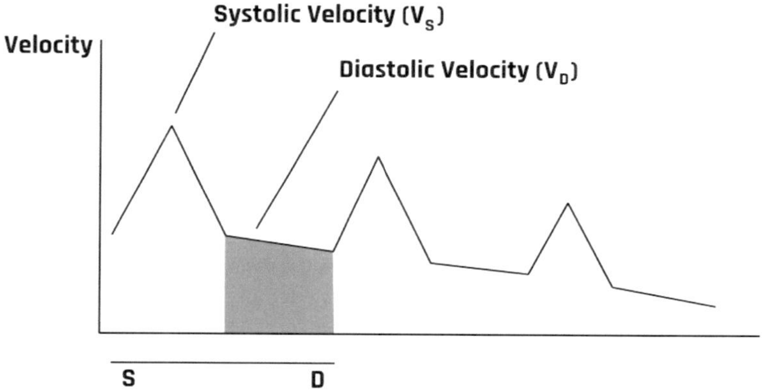

Figure 10-29: The transcranial Doppler velocity distribution.

A high PI (>1.2) can indicate increased intracranial pressure, microvascular disease, or distal vasospasm. Also, a low PI (<0.8) can be seen with carotid stenosis or occlusion as well as arteriovenous malformations.[7,8]

7. Davis D. Introduction to Transcranial Doppler Ultrasound [DVD]. St Petersburg (FL): Gulfcoast Ultrasound Institute; 2013 Feb 28–Mar 21.
8. Katz ML, Alexandrov AV. A Practical Guide to Transcranial Doppler Examinations. [place unknown]: Summer publishing; 2003. p. 158.

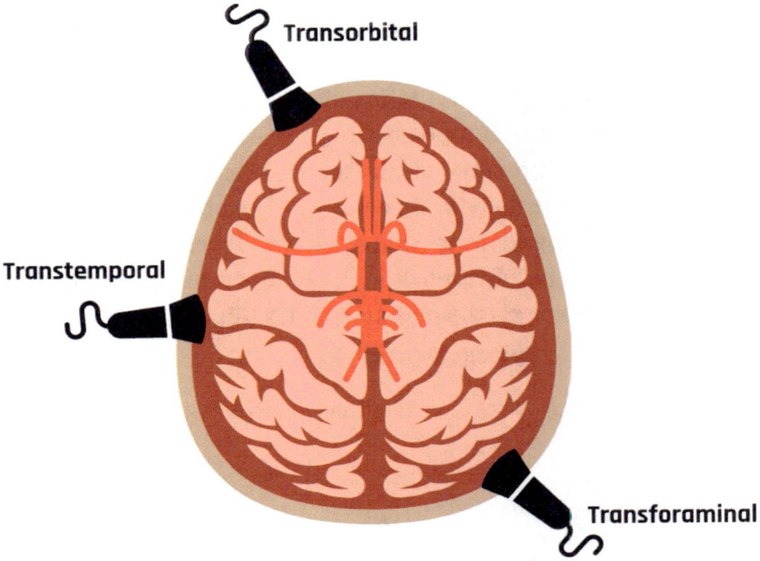

Figure 10-30: The three most common acoustic windows that provide direction to evaluate the intracranial vessels are the transtemporal, transorbital, and transforaminal windows.

The three most common acoustic windows that provide direction to evaluate the intracranial vessels are the transtemporal, transorbital, and transforaminal windows, as shown in Figure 10-30. The transcranial evaluation begins with the transtemporal approach on each side to identify the anterior, middle, and posterior cerebral arteries, and sometimes, the most distal aspect of the internal carotid artery (ICA) may also be evaluated.[9],[10]

9. Davis D. Introduction to Transcranial Doppler Ultrasound [DVD]. St Petersburg (FL): Gulfcoast Ultrasound Institute; 2013 Feb 28–Mar 21.
10. Katz ML, Alexandrov AV. A Practical Guide to Transcranial Doppler Examinations. [place unknown]: Summer publishing; 2003. p. 158.

When evaluating the anterior cerebral artery, normal flow is away from the probe, the depth is 60–70 mm, and the velocity ranges from 41–76 cm/sec. The middle cerebral artery has normal flow toward the probe, a depth of 30–60 mm, and a velocity that ranges from 46–86 cm/sec. The posterior cerebral artery typically has flow toward the probe, a depth of 60–70 mm, and a velocity range of 33–64 cm/sec, as shown in Figure 10-31.[11, 12, 13]

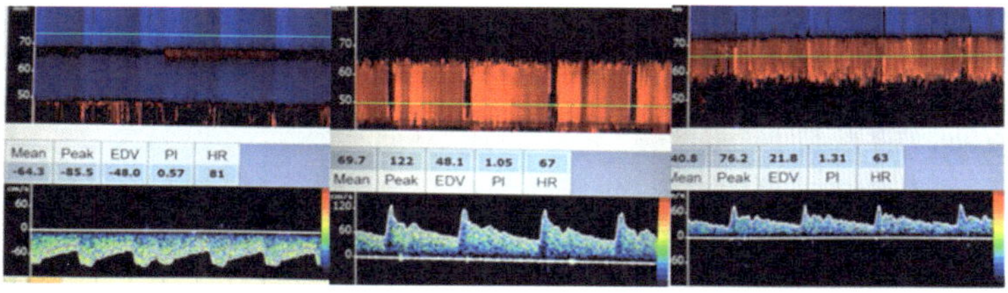

Figure 10-31: Side-by-side flow patterns and velocities of the anterior cerebral artery, middle cerebral artery, and posterior cerebral artery on transcranial Doppler.

The transorbital approach is followed and used to evaluate the ophthalmic artery and carotid siphon on each side, as shown in Figure 10-32. The location for of obtaining flow patterns is essential in this window, since it is difficult to determine the anatomic structure, as in the transtemporal approach demonstrating the circle of Willis. Comparisons can be made between different flow patterns.

11. Davis D. Introduction to Transcranial Doppler Ultrasound [DVD]. St Petersburg (FL): Gulfcoast Ultrasound Institute; 2013 Feb 28–Mar 21.
12. Katz ML, Alexandrov AV. A Practical Guide to Transcranial Doppler Examinations. [place unknown]: Summer publishing; 2003. p. 158.
13. Rumwell C. McPharlin M. Vascular Technology: An illustrated review. 4th ed. [place unknown]: Davies publishing; 2011. p. 442.

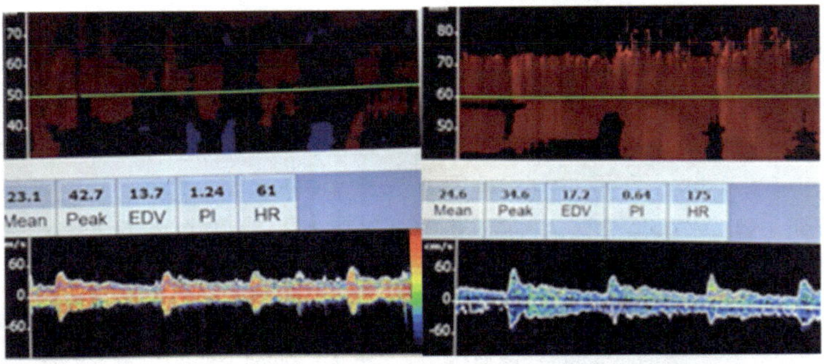

Figure 10-32: Side-by-side velocities and flow patterns of the right ophthalmic artery and carotid siphon on transcranial Doppler.

The transforaminal approach is then used to evaluate the intracranial vertebral arteries and the basilar arteries, as shown in Figure 10-33.

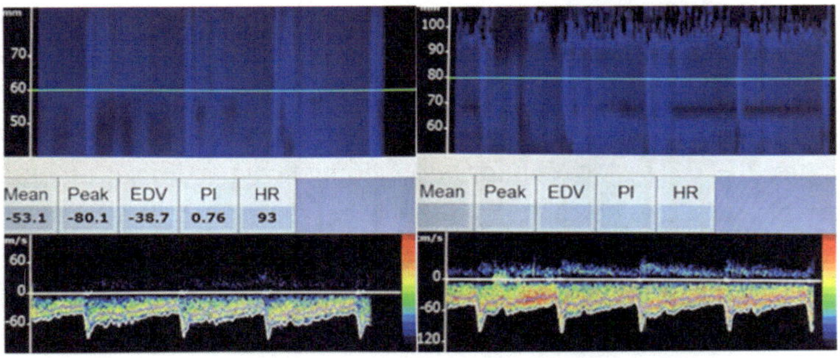

Figure 10-33: Side-by-side velocities and flow patterns of the left vertebral artery and basilar artery on transcranial Doppler.

In the vertebral arteries, normal blood flow is away from the probe, the depth is 60–70 mm, and the velocity ranges from 27–55 cm/sec. In the basilar arteries, normal blood flow is away from the probe, the depth is 80–120 mm, and the velocity is 30–57 cm/sec.

10.4.2 Sonography of the Carotid Arteries

Figure 10-34 shows the positioning of the ultrasound probe for carotid artery evaluations. A high-frequency linear transducer (7.5–10 MHz) is most appropriate for carotid sonography. Transverse and longitudinal views are both imaged in B-mode, color, and spectral Doppler. In the sagittal plane, the ICA, external carotid artery (ECA), and right common carotid artery (CCA) are followed from the clavicle to the mandible with anterior, oblique, lateral, and posterior projections to identify plaque formation. Comparison flow characteristics are made from one side to the other as well as from proximal to distal segments of the ICA, ECA, and CCA.[14] In the transverse plane, the ICA and CCA are followed to evaluate plaque formations. The percentage of stenosis can then be evaluated by looking at the diameter reduction. Plaque characteristics can be evaluated for calcification, thrombosis, and fibrosis.

14. Meissner MH, Moneta G, Burnand K, Gloviczki P, Lohr JM, Lurie F, Mattos MA, McLafferty RB, Mozes G, Rutherford RB, Padberg F, Sumner DS. The hemodynamics and diagnosis of venous disease. J Vasc Surg. 2007 Dec;46 Suppl S:4S–24S. doi: 10.1016/j.jvs.2007.09.043. PMID: 18068561.

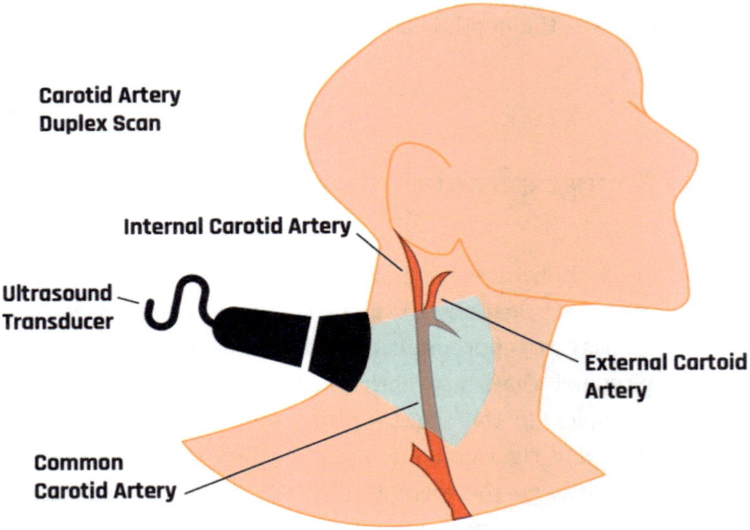

Figure 10-34: Carotid Doppler probe.

Figure 10-35 shows the CCA in the proximal transverse plane with the jugular vein on top, demonstrating blood flow in the opposite direction following the BART principle. *Prox* is the abbreviation used for *proximal* in the image for Figure 10-35 and some of the other following ones. Pulsed Doppler with spectral analysis is the primary tool for evaluating blood flow in the vascular study.

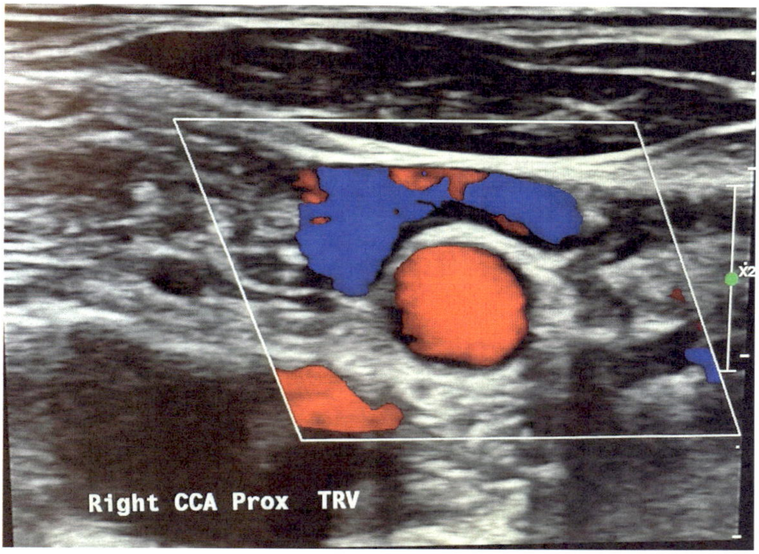

Figure 10-35: Right common carotid artery in the proximal transverse plane with the jugular vein on top, demonstrating blood flow in opposite directions.

Figure 10-36 shows the pulsed wave (PW) Doppler with spectral analysis of the right CCA in the distal sagittal plane. Spectral analysis is a method of displaying the variety of frequencies of blood flow during systole and diastole. The scanner technology automatically analyzes and displays the individual frequencies of the returned signals, creating a velocity profile consisting of time on the horizontal axis, frequency shifts on the vertical axis, and amplitude as brightness. This combination of blood flow analysis and anatomic information is the basis of duplex ultrasonography,[15] as discussed in Chapter 2.

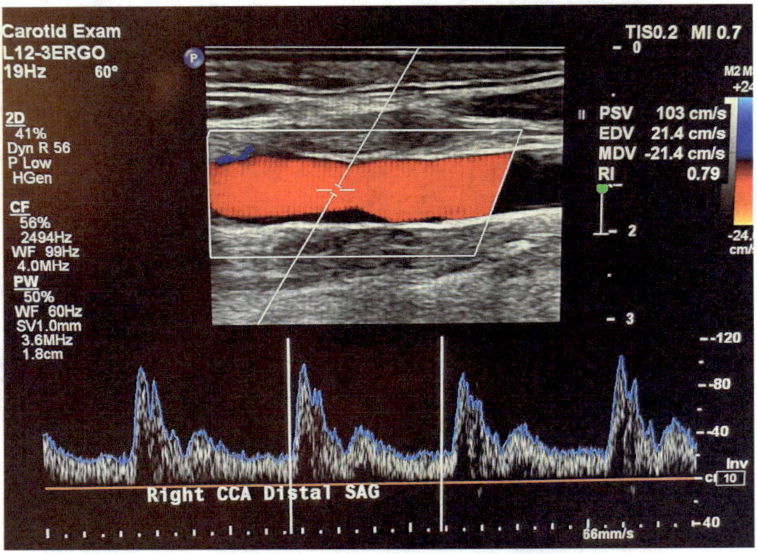

Figure 10-36: Pulsed wave Doppler with spectral analysis of the common carotid artery in the distal sagittal plane.

15. Rumwell C. McPharlin M. Vascular Technology: An illustrated review. 4th ed. [place unknown]: Davies publishing; 2011. p. 442.

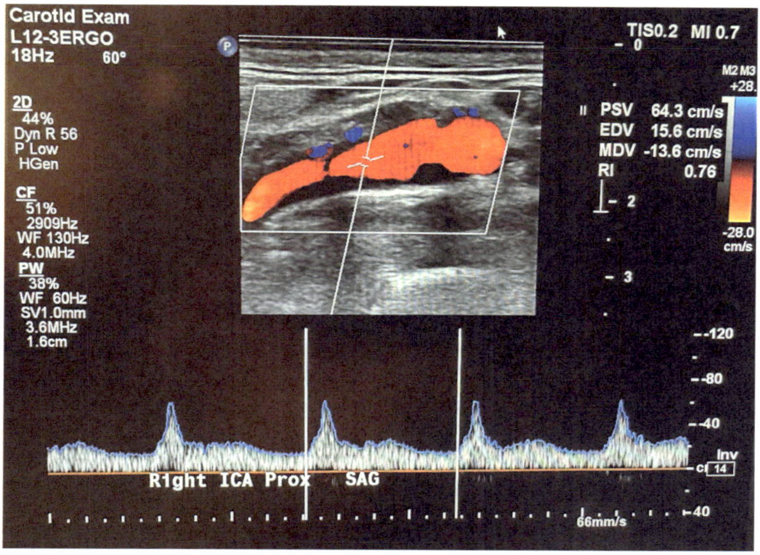

Figure 10-37: Normal waveform of the right internal carotid artery.

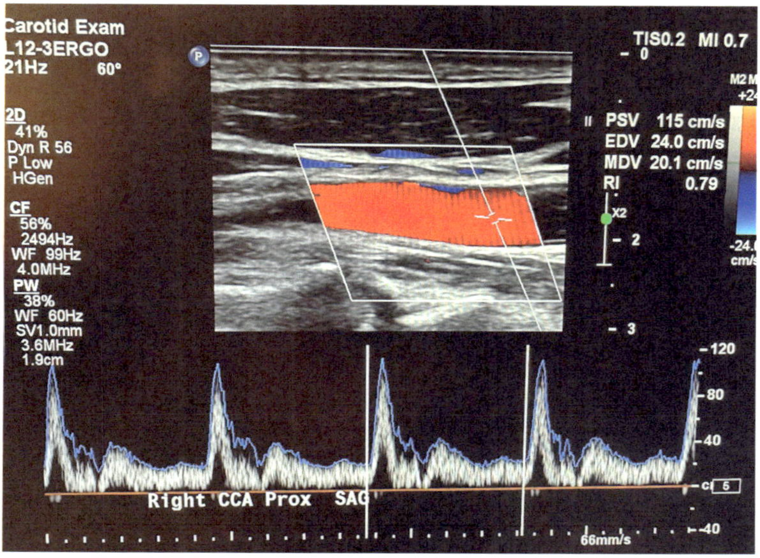

Figure 10-38: Normal waveform of the right common carotid artery.

The Doppler characteristics of the carotid artery system are different. The ICA and CCA usually have low flow resistance, as shown in Figures 10-37 and 10-38, respectively. Flow occurs throughout the cardiac cycle. The diastolic segment does not touch the baseline. The ECA has high flow resistance with little or no diastolic or reversed diastolic flow. Reproducible and consistent velocity measurements require an angle of 60 degrees or less. Although a zero-degree angle of insonation provides the most remarkable Doppler shift because this depends on the angle's cosine, this would be difficult with most vessels. The criteria used for the interpretation of velocity measurements were established using a 60-degree angle.[16, 17]

Since the brain is a low-resistance vascular bed, the ICA is less pulsatile with increased flow during diastole. The typical waveform of the ICA has a rapid upstroke during systole and a high diastolic component with a possible dicrotic notch and gradual downslope.

With the common carotid artery in the longitudinal plane, the transducer is angled more posterolaterally to identify the vertebral artery. Vertical shadows will appear running through the vertebral arteries, which are the transverse processes of the vertebrae, as shown in Figure 10-39. *Vert* in the image represents the vertebral artery. Flow direction is documented.

16. Green L, Jorgensen T, Schroedter B, Bendick P. Carotid Duplex and Color Flow Imaging [DVD]. St Petersburg (FL): Gulfcoast Ultrasound Institute; 2012 Feb 20–22.
17. Bandyk D, Armstrong PA, Neumyer MJ. Vascular Ultrasound Interpretation [DVD]. St Petersburg (FL): Gulfcoast Ultrasound Institute; 2012 August 9–10.

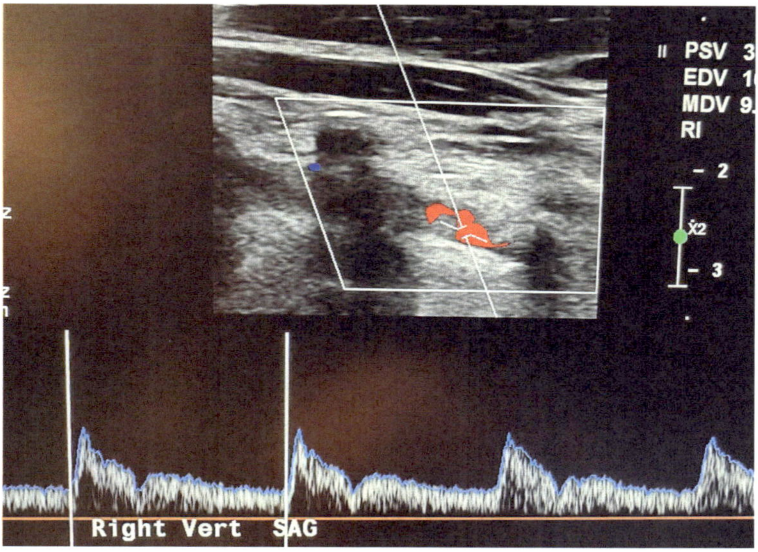

Figure 10-39: Sagittal view of the right vertebral artery with color flow.

The ECA supplies blood to vascular areas with higher resistance, such as the scalp. It has a rapid upstroke in systole and rapid downstroke in diastole with a dicrotic notch, as shown in Figure 10-40.

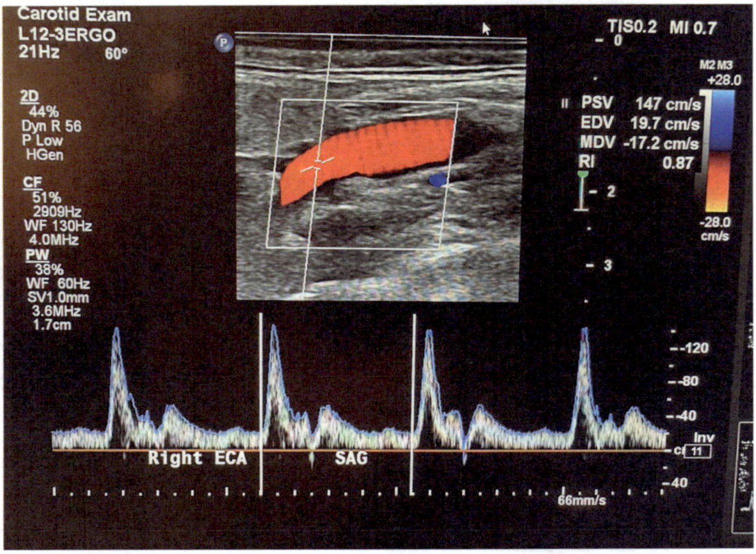

Figure 10-40: Normal waveform of the right external carotid artery.

Both the CCA and vertebral arteries have low flow resistance. The flow characteristics are similar to the ICA. Multiple guidelines and trials are used to determine the percentage of diameter stenosis and clinically relevant stenosis. The ICA is of the most importance for surgical intervention. The Society of Radiologists in Ultrasound Consensus, one of the most widely used guidelines to assess ICA stenosis, is presented in Table 10-1 the table below.[18]

18. Grant EG, Benson CB, Moneta GL, Alexandrov AV, Baker JD, Bluth EI, Carroll BA, Eliasziw M, Gocke J, Hertzberg BS, Katanick S, Needleman L, Pellerito J, Polak JF, Rholl KS, Wooster DL, Zierler RE. Carotid artery stenosis: Gray-scale and Doppler US diagnosis—Society of Radiologists in Ultrasound Consensus

Table 10-1: Criteria for assessing Internal Carotid Artery Stenosis.

	Primary Parameters		Additional Parameters	
Degree of Stenosis (%)	ICA PSV (cm/sec)	Plaque Estimate (%)*	ICA/CCA PSV Ratio	ICA EDV (cm/sec)
Normal	<125	None	<2.0	<40
<50	<125	<50	<2.0	<40
50–69	125–230	≥50	2.0–4.0	40–100
≥70 but less than near occlusion	>230	≥50	>4.0	>100
Near occlusion	High, low, or undetectable	Visible	Variable	Variable
Total occlusion	Undetectable	Visible, no detectable lumen	Not applicable	Not applicable

* Plaque estimate (diameter reduction) with gray-scale and color Doppler US.

The North American Symptomatic Endarterectomy Trial was published in the journal *Stroke* in 1991.[19] The results concluded that patients with 70–99% stenosis of the ICA benefit from surgical intervention in the appropriate clinical setting.

10.4.3 Sonography of the Aorta

The aorta is the largest artery in humans. It branches off the heart's left ventricle into the thoracic and abdominal cavities, as shown in Figure 10-41. The abdominal aorta branches into the right and left iliac arteries at the level of the umbilicus, where it carries oxygenated blood to each lower extremity. When the wall of the aorta weakens and expands, an aneurysm develops (with an increased risk of rupture under this high-pressure system).

Conference. Radiology. 2003 Nov;229(2):340–6. doi: 10.1148/radiol.2292030516. Epub 2003 Sep 18. PMID: 14500855.

19. North American Symptomatic Carotid Endarterectomy Trial. Methods, patient characteristics, and progress. Stroke. 1991 Jun;22(6):711–20. doi: 10.1161/01.str.22.6.711. PMID: 2057968.

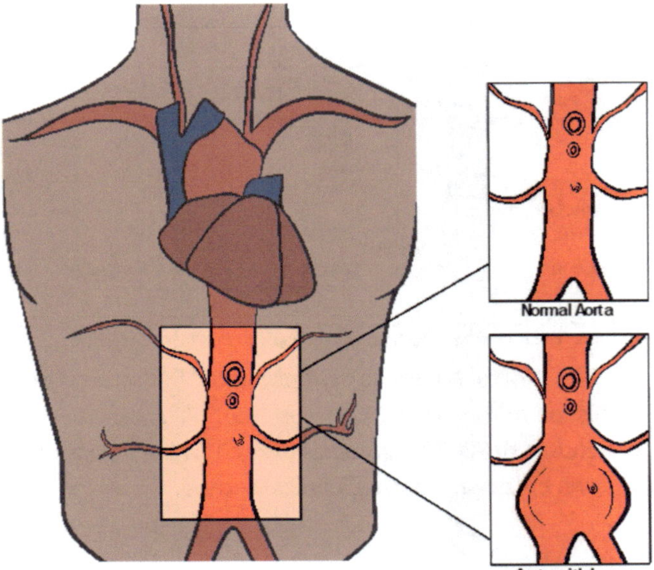

Figure 10-41: Schematic of the abdominal aorta showing that it branches into the right and left iliac arteries.

Each year, 200,000 people in the United States are diagnosed with an abdominal aortic aneurysm (AAA). Of these, nearly 7.5% have a life-threatening risk of rupture. The majority of patients with AAA are asymptomatic. The aorta's diameter must be greater less than 3 cm to be diagnosed as an aneurysm. When it reaches 5 cm or greater, very close monitoring and surgical options are entertained. Risk factors for an AAA include hypertension, smoking, and genetic factors, especially involving immediate relatives with AAA. Men over the age of 60 are also at greater risk.[20]

20. Chaikof EL, Dalman RL, Eskandari MK, Jackson BM, Lee WA,

The ultrasound evaluation of the abdominal aorta (A or AA) should include the proximal, mid, and distal aorta to the bifurcation in the transverse and longitudinal planes, as shown in Figure 10-42. Evaluation of the branches of the aorta should include the celiac artery (C), superior mesenteric artery (SMA), and renal artery branches, as shown in Figures 10-43 and 10-44, without and with color Doppler, respectively.[21] The abbreviations given in parentheses in this paragraph have been labeled in some of the ultrasound images discussed below.

Mansour MA, Mastracci TM, Mell M, Murad MH, Nguyen LL, Oderich GS, Patel MS, Schermerhorn ML, Starnes BW. The Society for Vascular Surgery practice guidelines on the care of patients with an abdominal aortic aneurysm. J Vasc Surg. 2018 Jan;67(1):2–77.e2. doi: 10.1016/j.jvs.2017.10.044. PMID: 29268916.

21. Meissner MH, Moneta G, Burnand K, Gloviczki P, Lohr JM, Lurie F, Mattos MA, McLafferty RB, Mozes G, Rutherford RB, Padberg F, Sumner DS. The hemodynamics and diagnosis of venous disease. J Vasc Surg. 2007 Dec;46 Suppl S:4S–24S. doi: 10.1016/j.jvs.2007.09.043. PMID: 18068561.

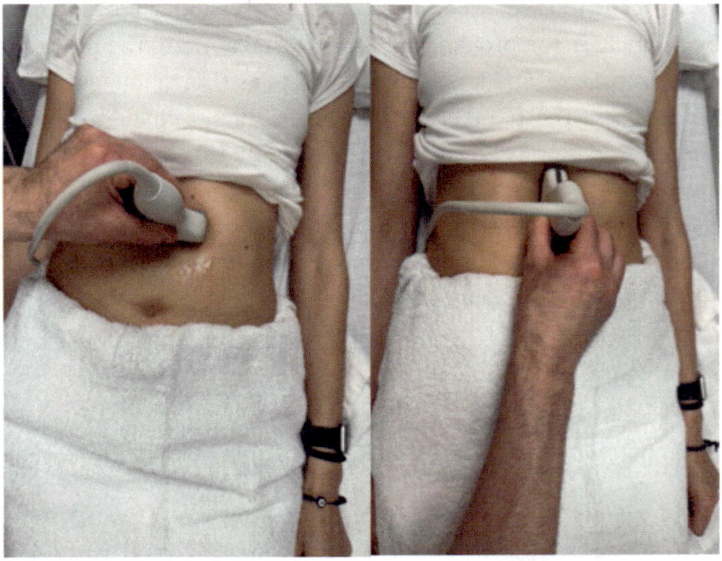

Figure 10-42: Transducer position in the transverse and sagittal planes to evaluate the abdominal aorta.

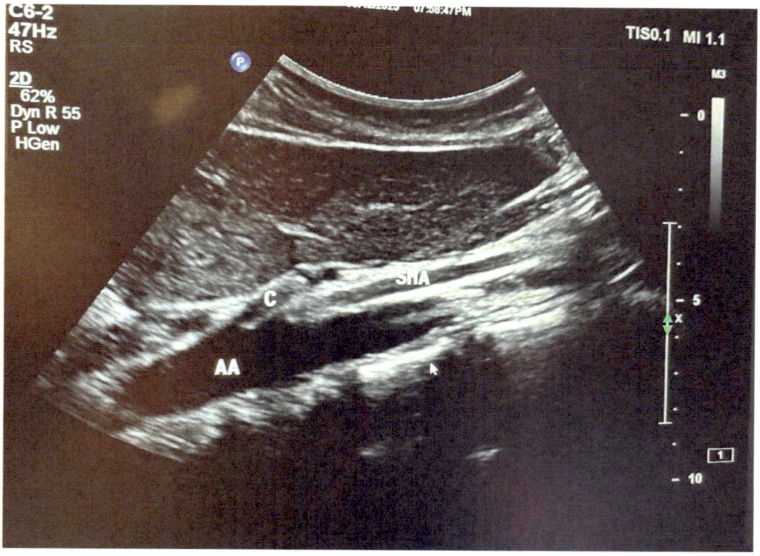

Figure 10-43: Sagittal image of proximal aorta without Doppler showing the abdominal aorta, celiac artery, and superior mesenteric artery.

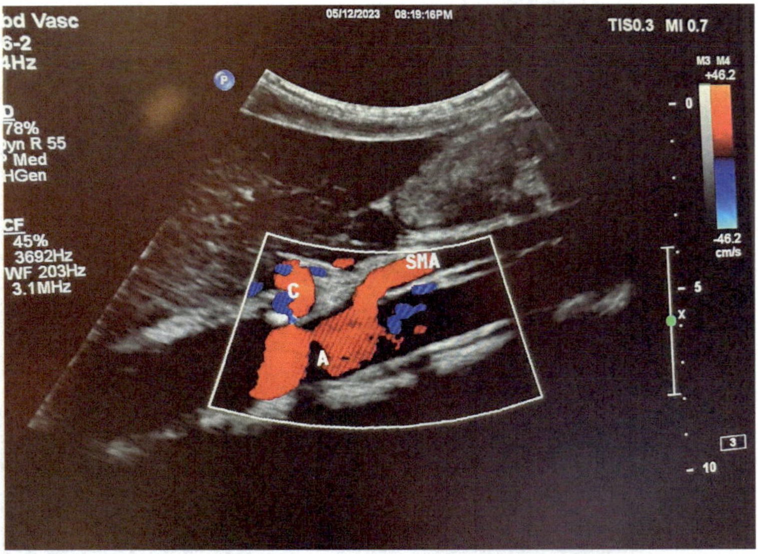

Figure 10-44: Sagittal image with color Doppler showing the abdominal aorta, celiac artery, and superior mesenteric artery.

10.5 Arterial Sonography of the Lower Extremities

Figure 10-45 shows the normal anatomical branches of the arterial system in the lower extremity. Peripheral artery disease (PAD) is a condition in which the arteries of the lower extremities are narrowed primarily from atherosclerosis. Approximately 8 million people in the United States have PAD. Men and women are affected equally. Risk factors include smoking, diabetes, hypertension, high cholesterol, and being over 60 years of age. A classic symptom of PAD is claudication, or pain when walking. Lower-extremity arterial duplex scanning is a noninvasive way to identify the presence and severity of arterial occlusive disease.

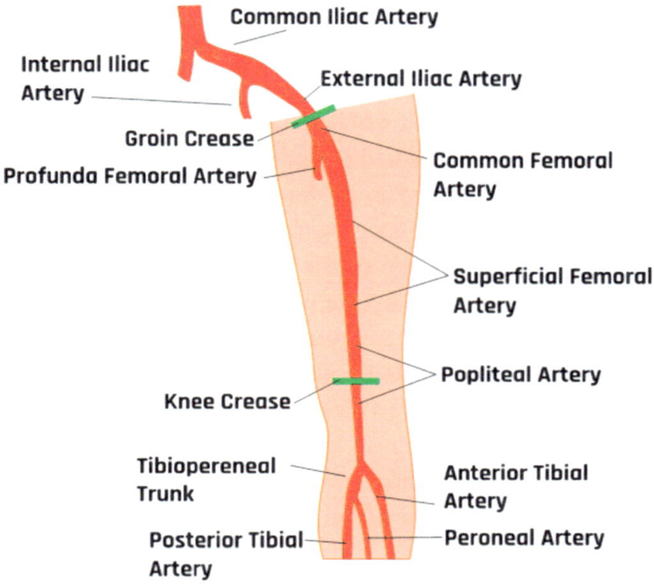

Figure 10-45: Schematic showing normal anatomical branches of the arterial system in the lower extremity.

It can also be used to follow the progression of the disease. The patient must rest for at least 20 minutes before testing, since this can affect the results, especially if the patient has PAD. The patient is then positioned supine with the lower extremities at the heart level so the hydrostatic pressure cannot falsely elevate the measurements.

During an arterial Doppler exam, various cuffs are placed on the patient's legs and arms. This exam uses color wave (CW) Doppler. CW Doppler employs two crystals contained in the same probe: one that transmits the signal and one that receives the reflected sound wave of the blood cells. The reflected frequency is either higher or lower than the transmitted frequency, depending on the flow direction. This change in frequency is called the Doppler shift. The ankle brachial index (ABI) is recorded, and the waveforms are analyzed. The ABI is a simple test that compares the blood pressure in the upper and lower limbs. The ABI is calculated by dividing the blood pressure in an ankle artery by the blood pressure in an arm artery. An ABI value of less than 0.9 indicates PAD.

An arterial duplex is another type of evaluation that uses ultrasound. It starts proximally at the common femoral artery with a side-by-side transverse image without and with color Doppler, followed by a sagittal image of the artery in red and sometimes the corresponding vein(s) in blue, and finally, a sagittal image of the artery with waveform analysis, which includes peak systolic velocity (PSV) and end-diastolic velocity (EDV). As the arterial study is performed from proximal to distal, the same approach is obtained with each artery, including, in succession, the common femoral artery (CFA), profunda femoral artery (Prof A), superficial femoral artery (SFA), popliteal artery (Pop A), posterior tibial artery (PTA), peroneal artery (Pero A), anterior tibial artery (ATA), and dorsalis pedis artery (DPA). The abbreviations given in parentheses in the previous sentence have been labeled in some of the ultrasound images discussed below. The abbreviation *Trans* used in some of these images is for *transverse*. Figures 10-46 and 10-47 show the transverse and sagittal views, respectively, of the side-by-side images of the right common femoral artery without and with color Doppler. Figure 10-48 shows the sagittal view of the right common femoral artery with color Doppler and waveform analysis.

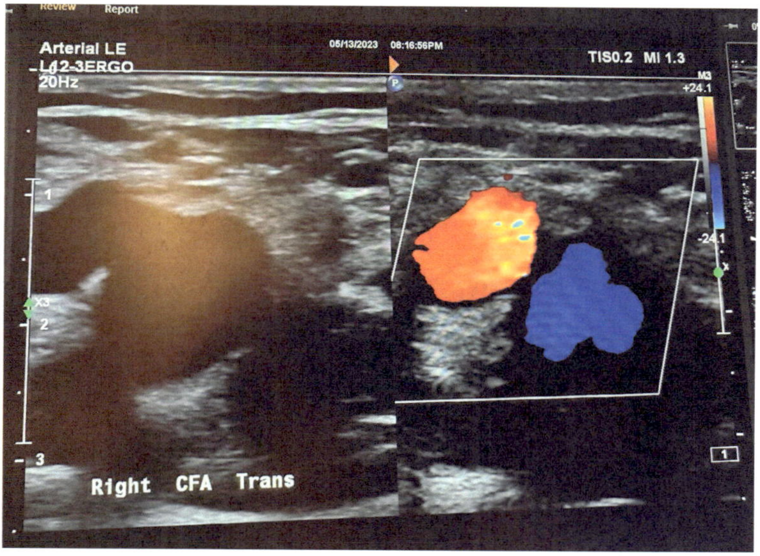

Figure 10-46: Side-by-side transverse images of the right common femoral artery without color Doppler and with color Doppler.

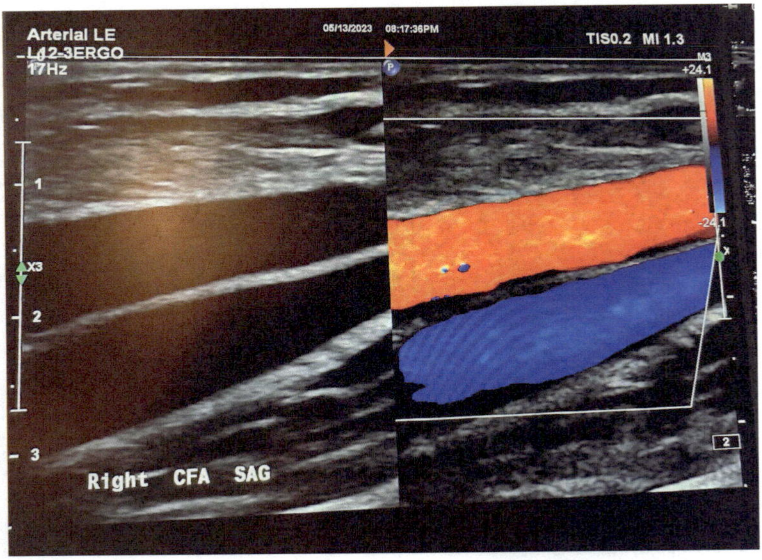

Figure 10-47: Side-by-side sagittal images of the right common femoral artery without color Doppler and with color Doppler.

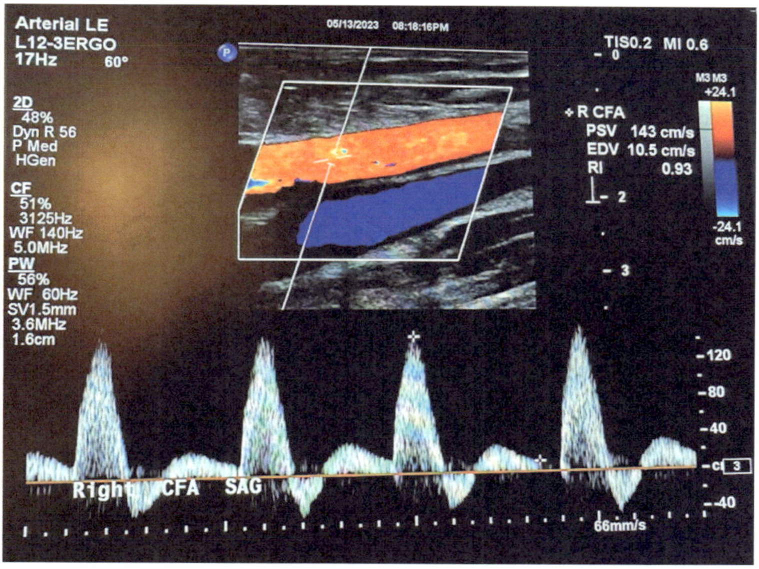

Figure 10-48: Sagittal image of the right common femoral artery with color Doppler and waveform analysis.

Figures 10-49 and 10-50 show the transverse and sagittal views, respectively, of the right profunda femoral artery without and with color Doppler. Figure 10-51 shows the sagittal view of the right profunda femoral artery with color Doppler and waveform analysis.

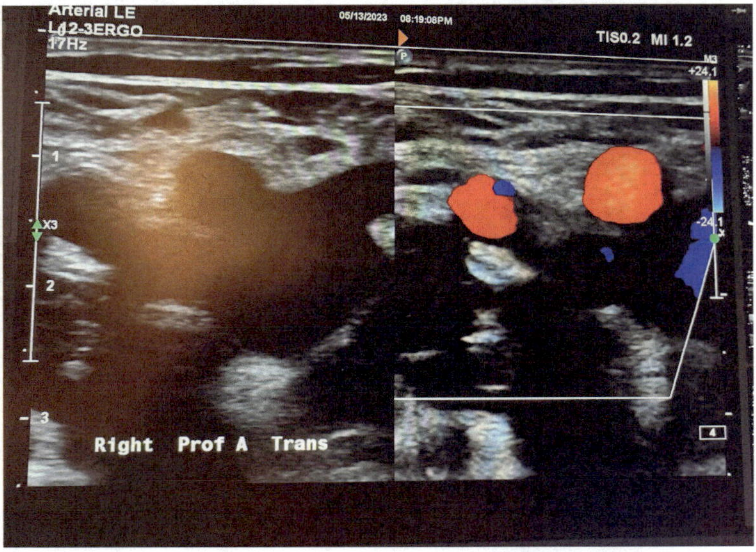

Figure 10-49: Side-by-side transverse images of the right profunda femoral artery without color Doppler and with color Doppler.

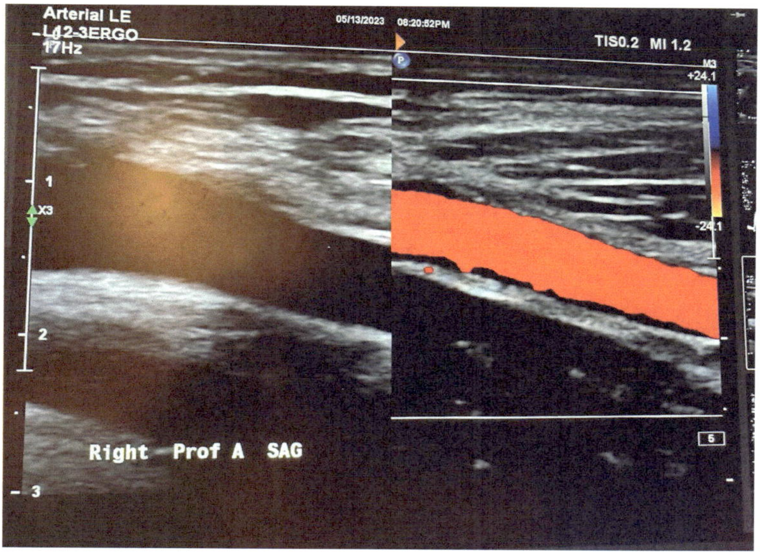

Figure 10-50: Side-by-side sagittal images of the right profunda femoral artery without color Doppler and with color Doppler.

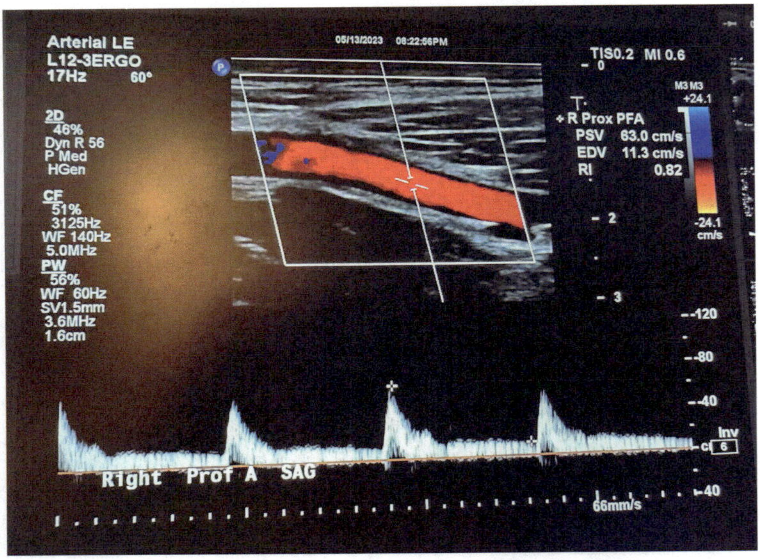

Figure 10-51: Sagittal image of the right profunda femoral artery with color Doppler and waveform analysis.

Figures 10-52 and 10-53 show the transverse and sagittal views, respectively, of the right proximal superficial femoral artery without and with color Doppler. Figure 10-54 shows the sagittal view of the right proximal superficial femoral artery with color Doppler and waveform analysis.

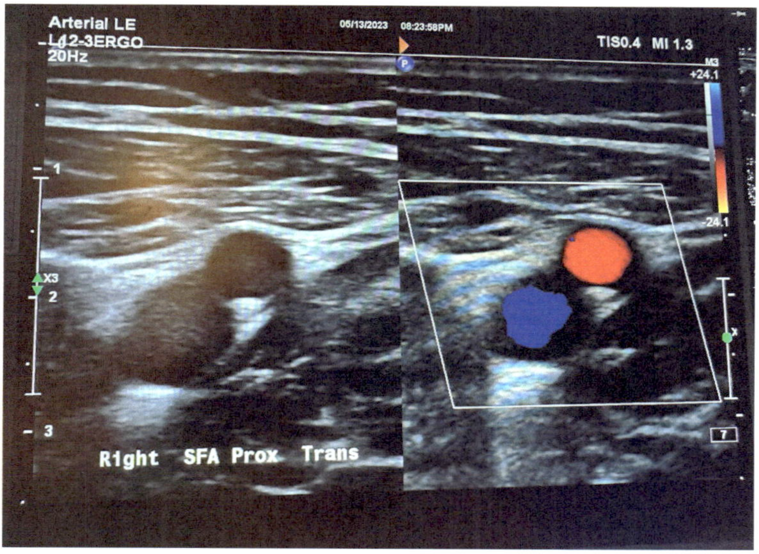

Figure 10-52: Side-by-side transverse images of the right proximal superficial femoral artery without color Doppler and with color Doppler.

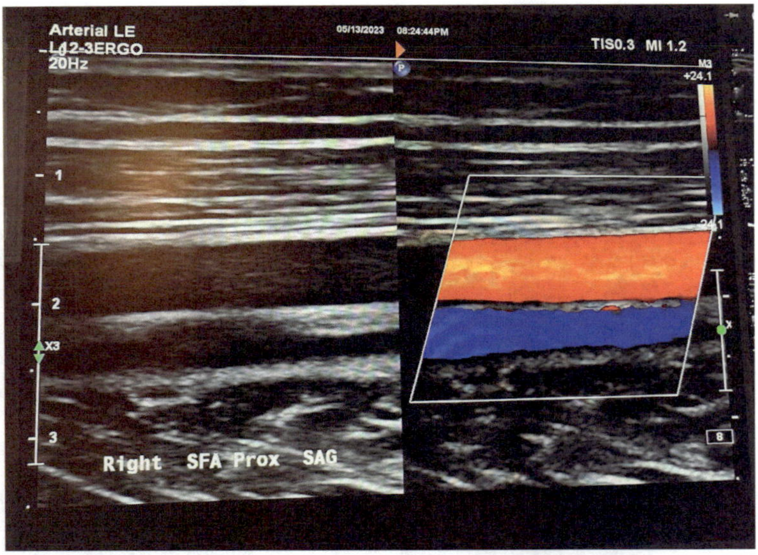

Figure 10-53: Side-by-side sagittal images of the right proximal superficial femoral artery without color Doppler and with color Doppler.

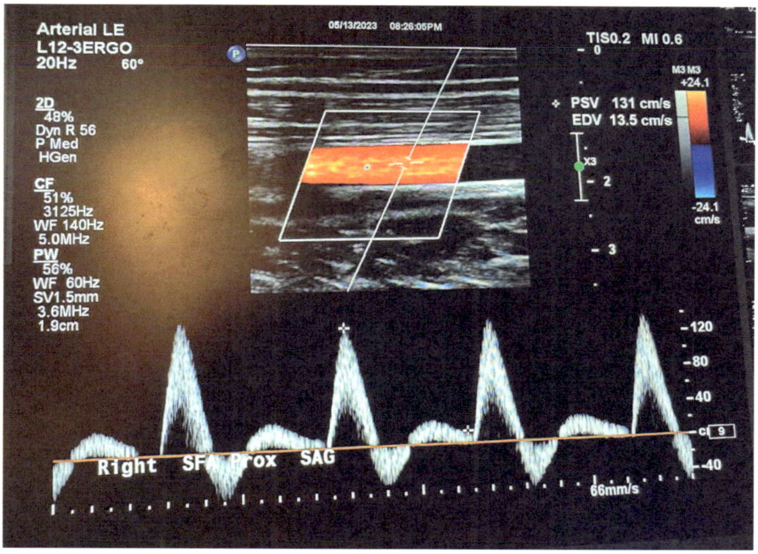

Figure 10-54: Sagittal image of the right proximal superficial femoral artery with color Doppler and waveform analysis.

Figures 10-55 and 10-56 show the transverse and sagittal views, respectively, of the right middle superficial femoral artery without and with color Doppler. Figure 10-57 shows the sagittal view of the right middle superficial femoral artery with color Doppler and waveform analysis.

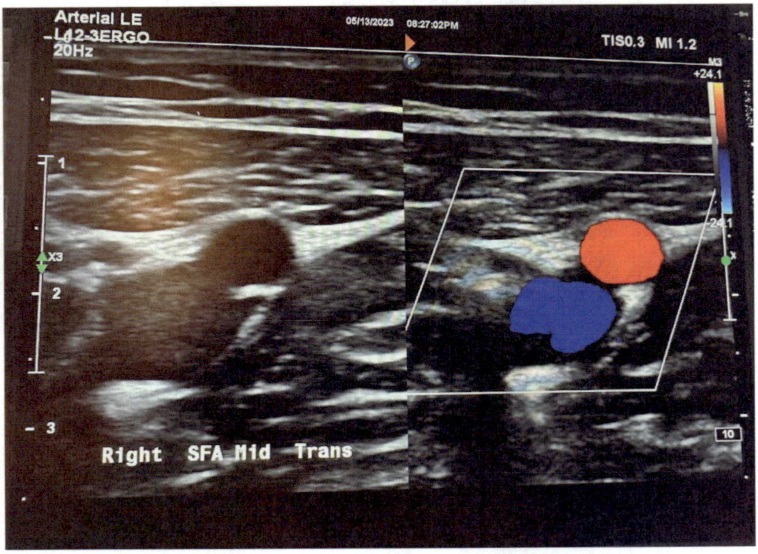

Figure 10-55: Side-by-side transverse images of the right middle superficial femoral artery without color Doppler and with color Doppler.

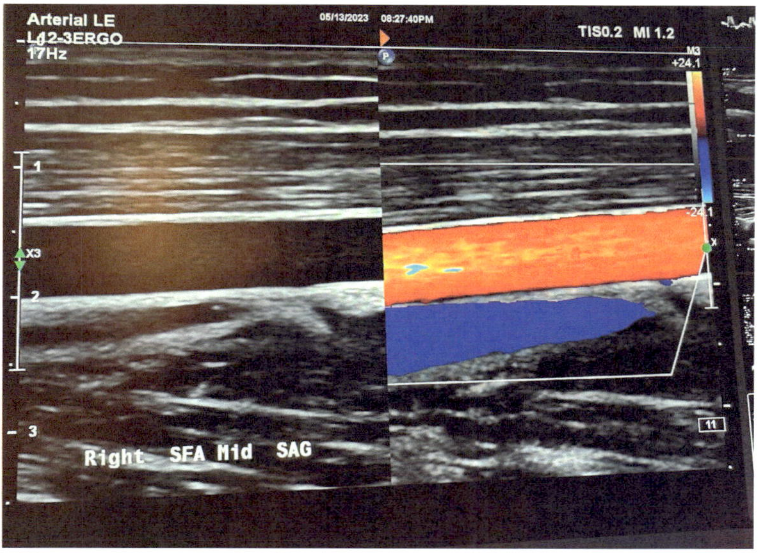

Figure 10-56: Sagittal image of the right middle superficial femoral artery without color Doppler and with color Doppler.

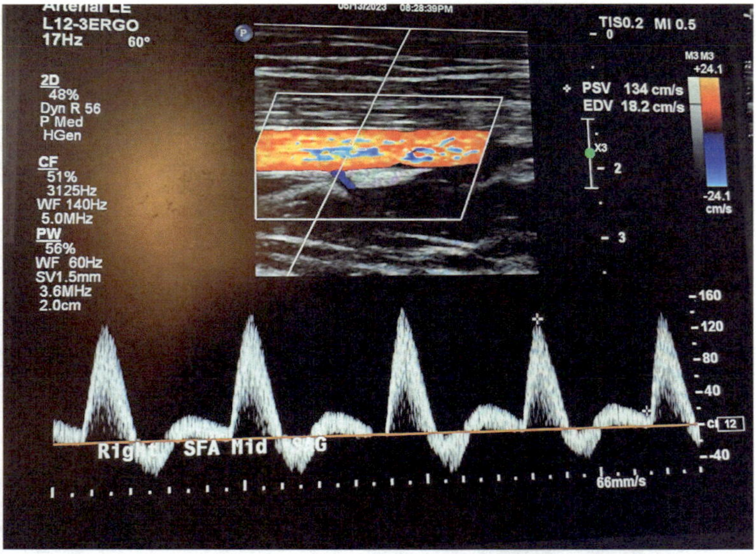

Figure 10-57: Sagittal image of the right middle superficial femoral artery with color Doppler and waveform analysis.

Figures 10-58 and 10-59 show the transverse and sagittal views of the right distal superficial femoral artery without and with color Doppler. Figure 10-60 shows the sagittal view of the right distal superficial femoral artery with color Doppler and waveform analysis.

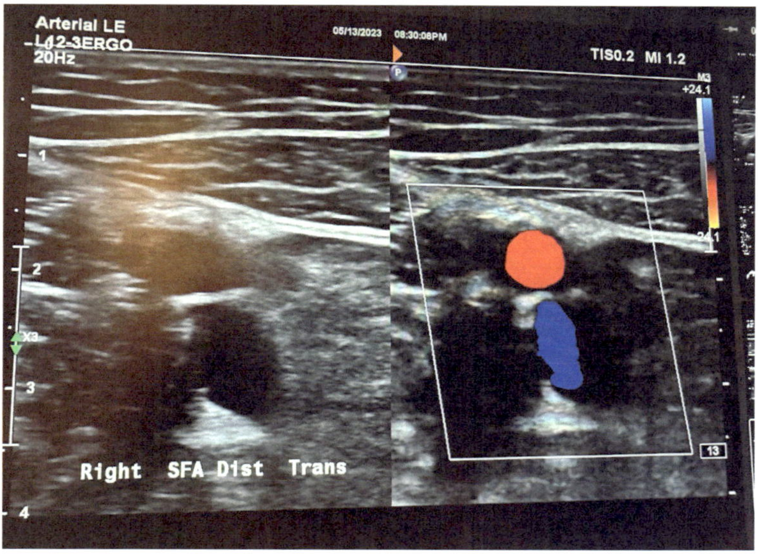

Figure 10-58: Side-by-side transverse images of the right distal superficial femoral artery without color Doppler and with color Doppler.

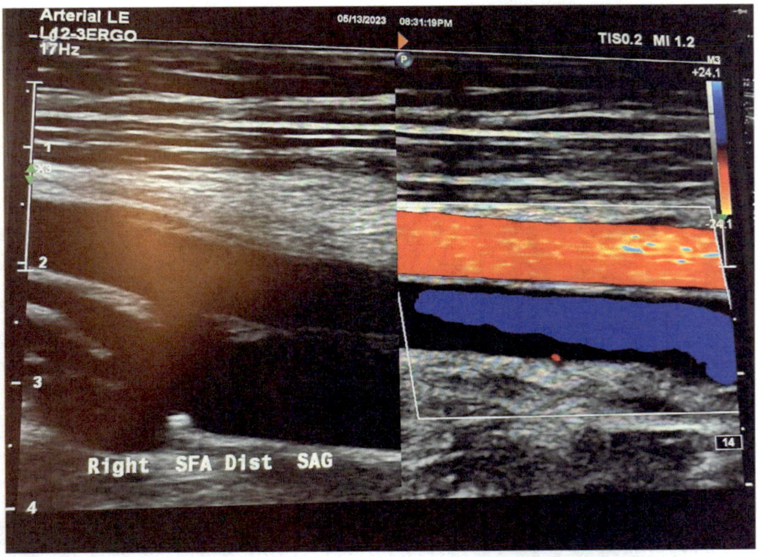

Figure 10-59: Sagittal images of the right distal superficial femoral artery without color Doppler and with color Doppler.

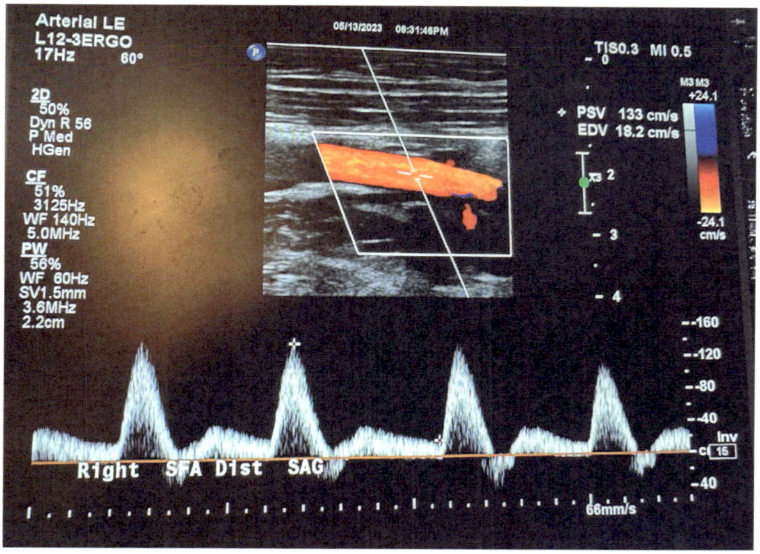

Figure 10-60: Sagittal image of the right distal superficial femoral artery with color Doppler and waveform analysis.

Figures 10-61 and 10-62 show the transverse and sagittal views, respectively, of the right popliteal artery without and with color Doppler. Figure 10-63 shows the sagittal view of the right popliteal artery with color Doppler and waveform analysis.

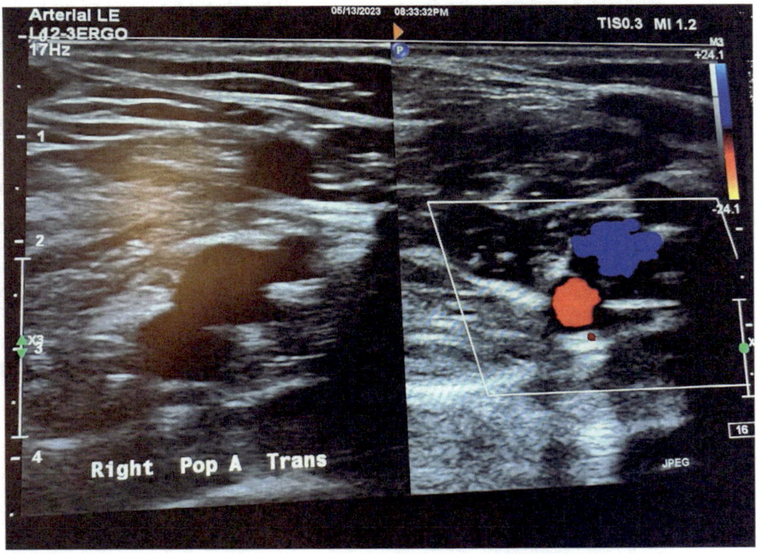

Figure 10-61: Side-by-side transverse images of the right popliteal artery without color Doppler and with color Doppler.

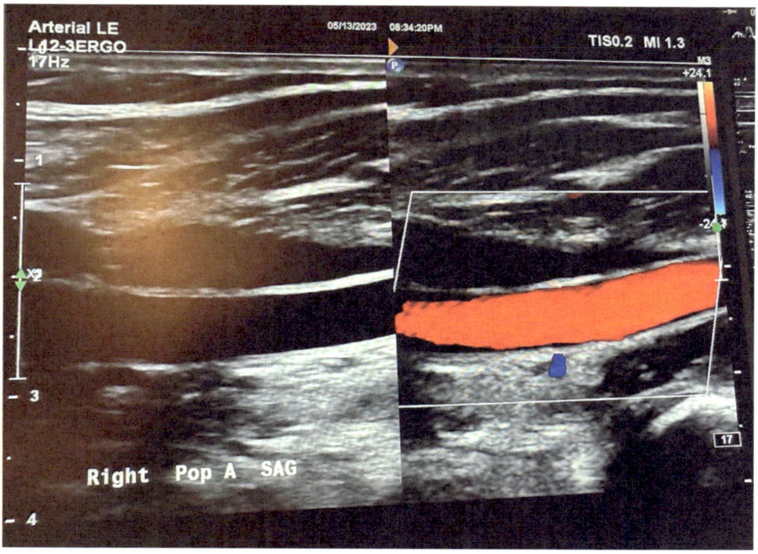

Figure 10-62: Side-by-side sagittal images of the right popliteal artery without color Doppler and with color Doppler.

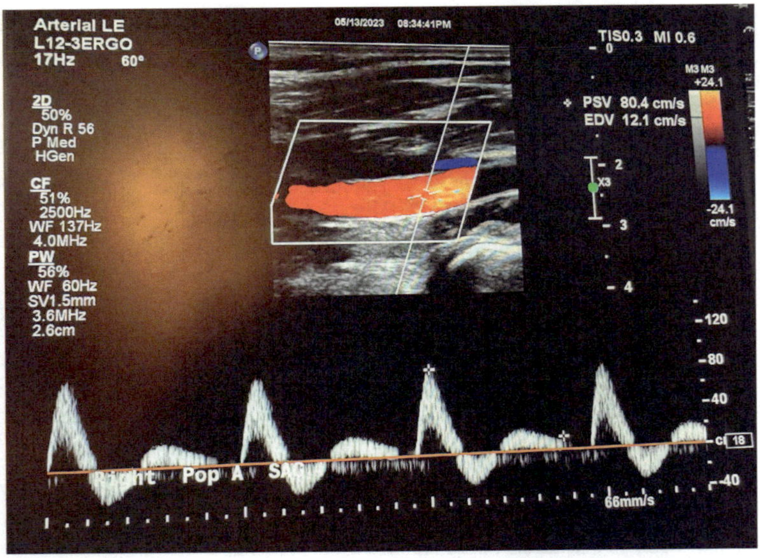

Figure 10-63: Sagittal image of the right popliteal artery with color Doppler and waveform analysis.

Figures 10-64 and 10-65 show the transverse and sagittal views, respectively, of the right posterior tibial artery without and with color Doppler. Figure 10-66 shows the sagittal view of the right posterior tibial artery with color Doppler and waveform analysis.

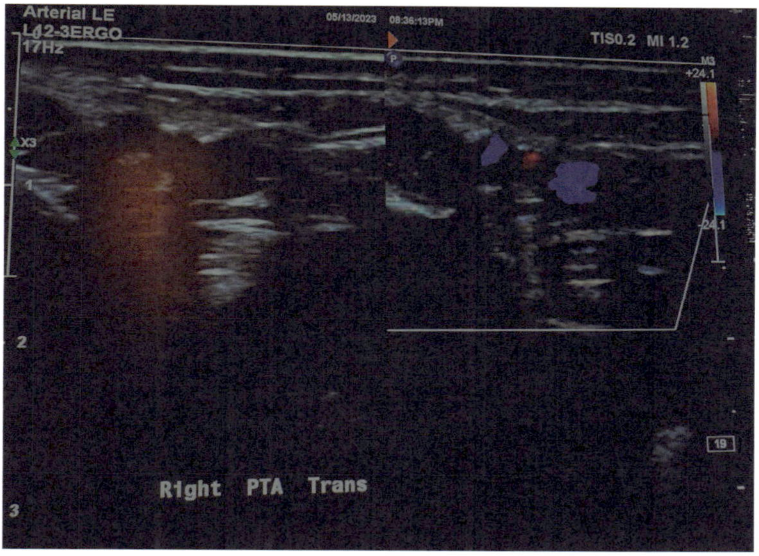

Figure 10-64: Side-by-side transverse images of the right posterior tibial artery without color Doppler and with color Doppler.

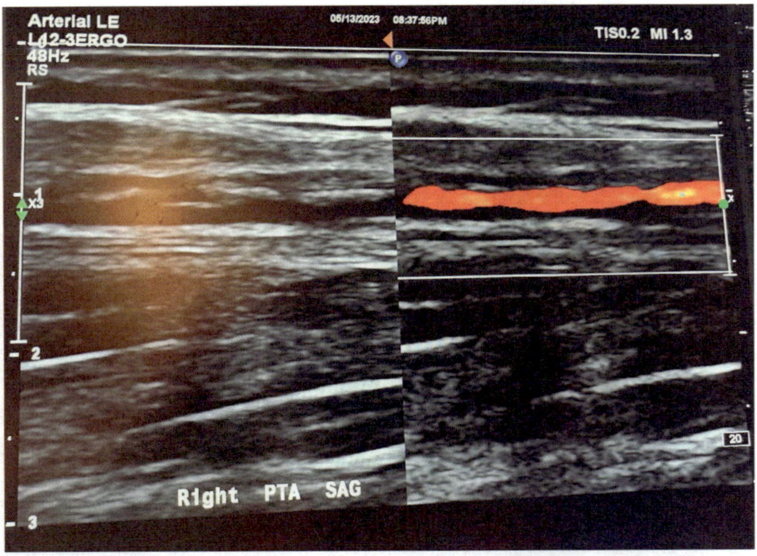

Figure 10-65: Side-by-side sagittal images of the right posterior tibial artery without color Doppler and with color Doppler.

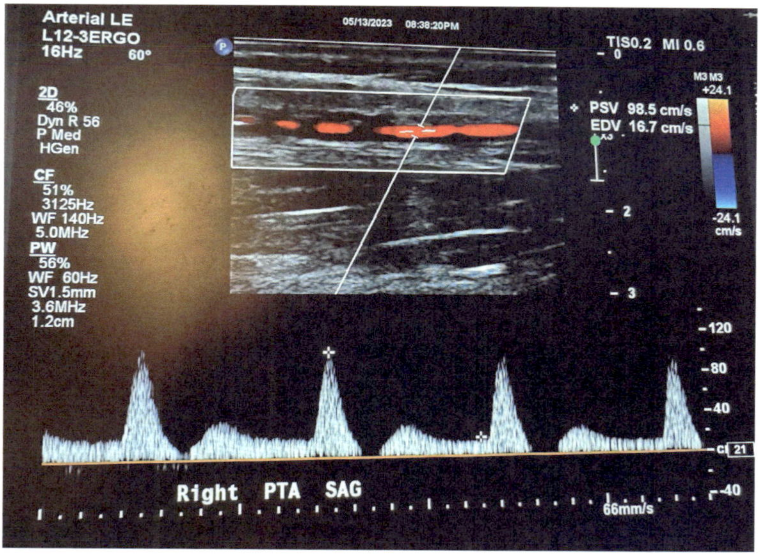

Figure 10-66: Sagittal image of the right posterior tibial artery with color Doppler and waveform analysis.

Figures 10-67 and 10-68 show the transverse and sagittal views, respectively, of the right peroneal artery without and with color Doppler. Figure 10-69 shows the sagittal view of the right peroneal artery with color Doppler and waveform analysis.

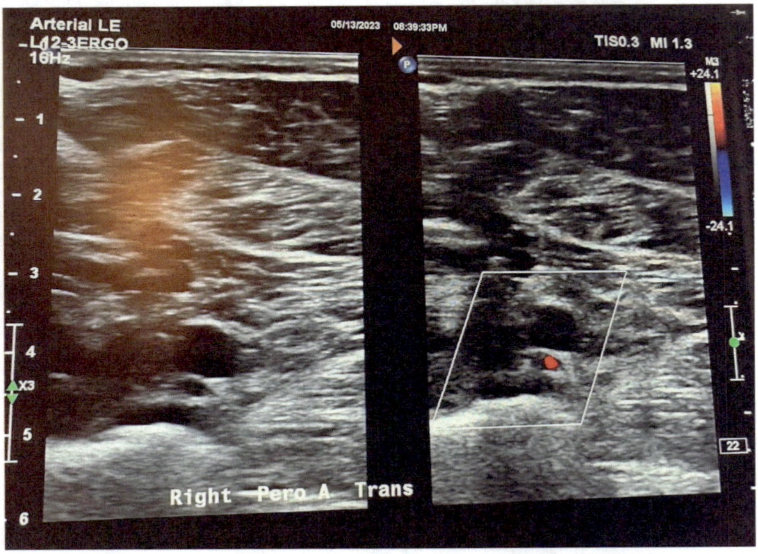

Figure 10-67: Side-by-side transverse images of the right peroneal artery without color Doppler and with color Doppler.

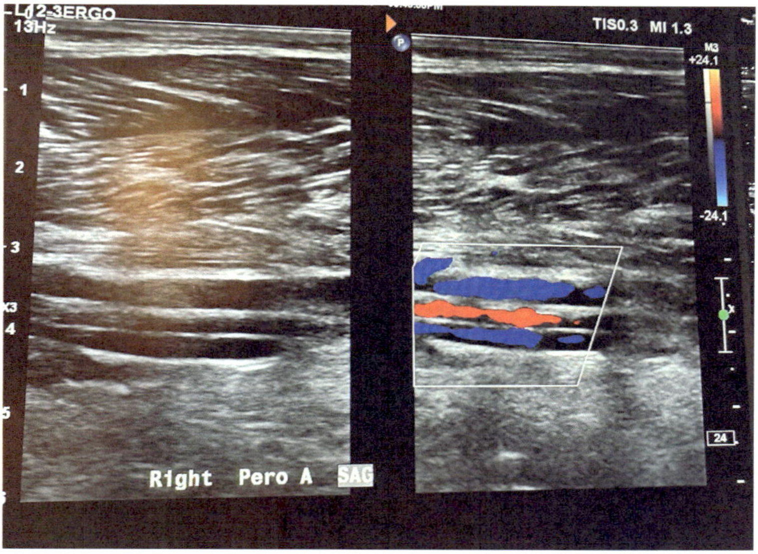

Figure 10-68: Side-by-side sagittal images of the right peroneal artery without color Doppler and with color Doppler.

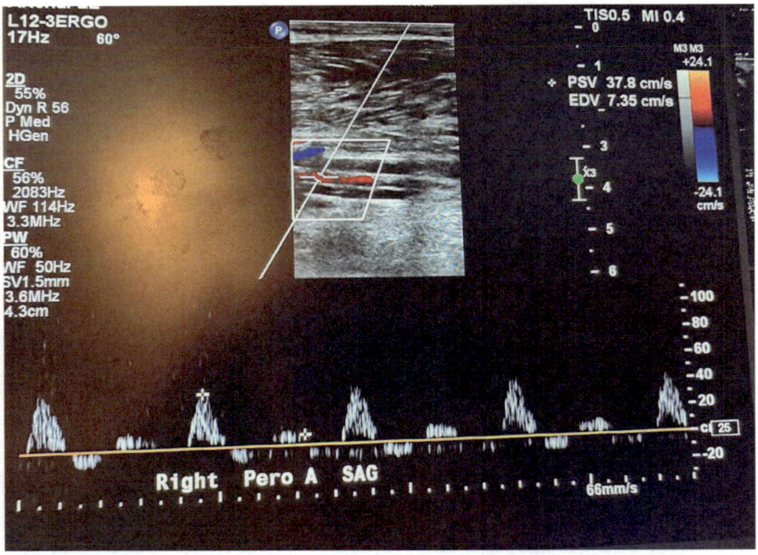

Figure 10-69: Sagittal image of the right peroneal artery with color Doppler and waveform analysis.

Figure 10-70 shows the side-by-side sagittal view of the right anterior tibial artery without and with color Doppler. Figure 10-71 shows the sagittal view of the right anterior tibial artery with color Doppler and waveform analysis.

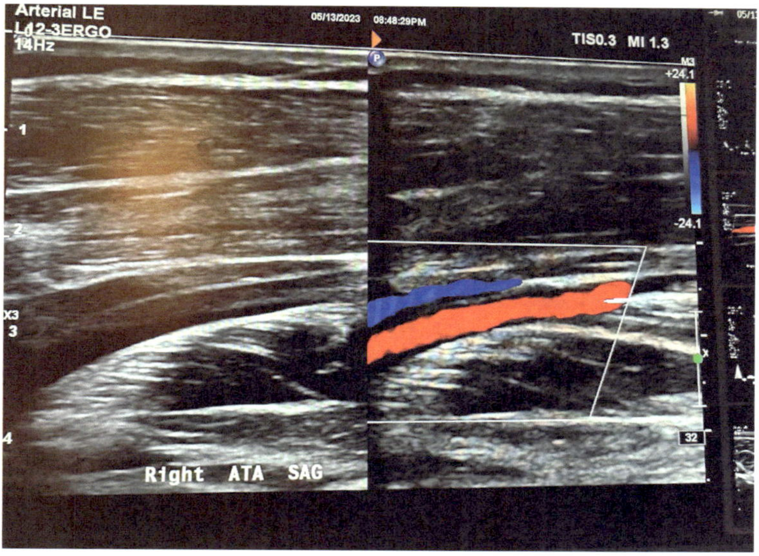

Figure 10-70: Side-by-side sagittal images of the right anterior tibial artery without color Doppler and with color Doppler.

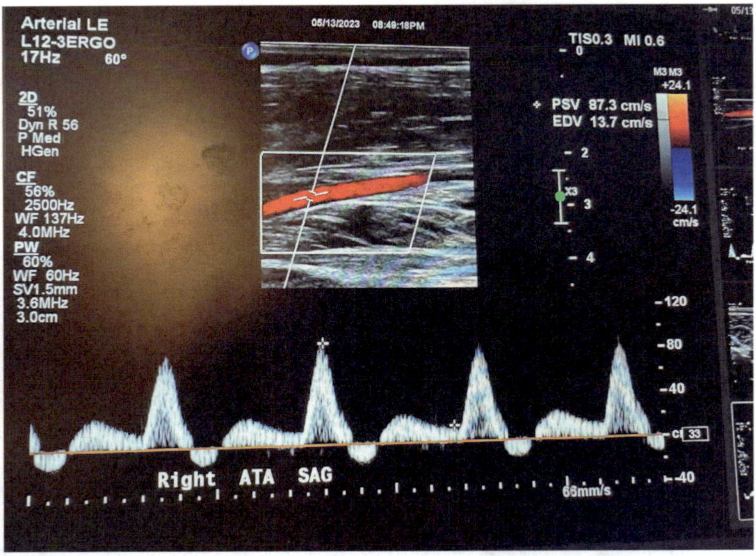

Figure 10-71: Sagittal image of the right anterior tibial artery with color Doppler and waveform analysis.

Figure 10-72 shows the transverse view of the right dorsalis pedis artery without and with color Doppler. Figure 10-73 shows the sagittal view of the right dorsalis pedis artery with color Doppler and waveform analysis.

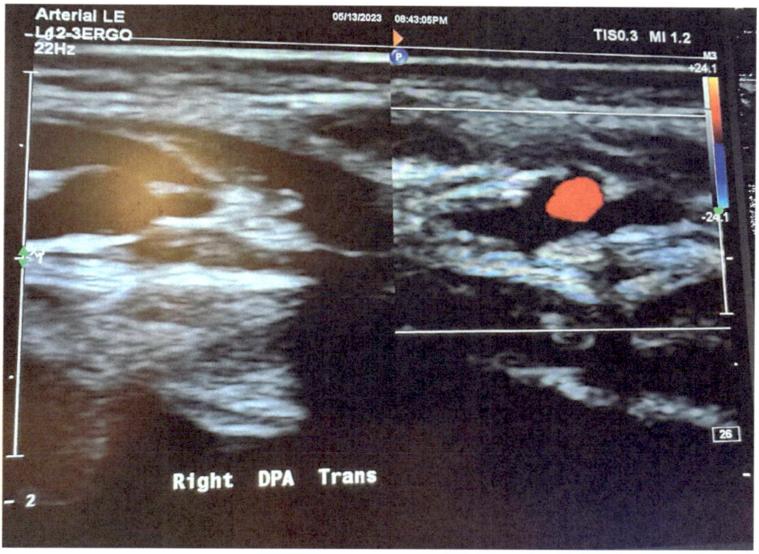

Figure 10-72: Side-by-side transverse images of the right dorsalis pedis artery without color Doppler and with color Doppler.

Figure 10-73: Sagittal image of the right dorsalis pedis artery with color Doppler.

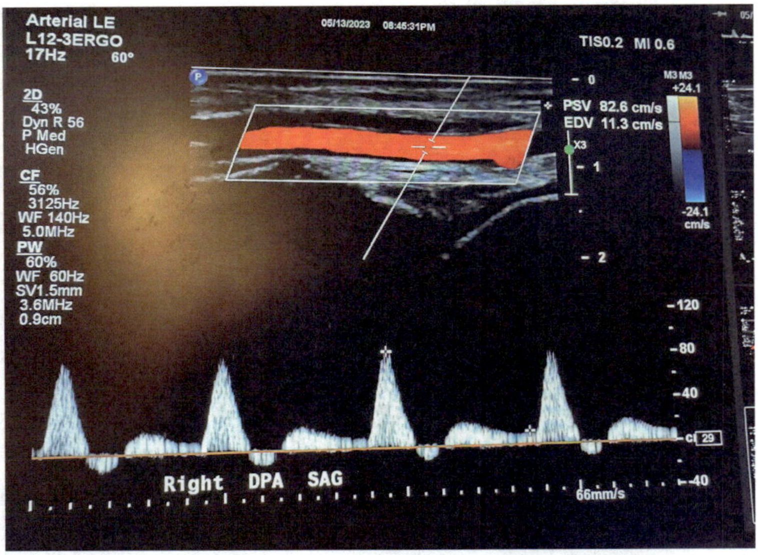

Figure 10-73: Sagittal image of the right dorsalis pedis artery with color Doppler and waveform analysis.

When evaluating the lower extremities for diagnostic criteria for PAD, the PSV and velocity ratio (VR) are often used. The VR is defined as the ratio of the PSV of the stenotic area to the PSV of the standard proximal segment. The degree of stenosis is determined by these values of the PSV and VR, as illustrated in Table 10-2.[22]

22. Hodgkiss-Harlow KD, Bandyk DF. Interpretation of arterial duplex testing of lower-extremity arteries and interventions. Semin Vasc Surg. 2013 Jun–Sep;26(2–3):95–104. doi: 10.1053/j.semvascsurg.2013.11.002. Epub 2013 Nov 14. PMID: 24636606.

Table 10-2: Guidelines for determining the degree of stenosis.

Degree of stenosis	Peak systolic velocity (cm/s)	Velocity ratio
<20%	<150	<1.5
20-49%	150-200	1.5-2.0
50-80%	200-300	2.0-4.0
>80%	>300	>4.0
Occlusion	No flow detected in lumen	N/A

10.6 Self-Assessment

1. What happens to ultrasound imaging of normal veins when you apply downward pressure with the transducer?
2. What frequency transducer would you typically use to evaluate the circle of Willis?
3. What kind of Doppler is used to evaluate the arterial system of the lower extremities?

10.7 *Further Readings*

1. Pugsley MK, Tabrizchi R. The vascular system. An overview of structure and function. J Pharmacol Toxicol Methods. 2000 Sep–Oct;44(2):333–40. doi: 10.1016/s1056-8719(00)00125-8. PMID: 11325577.
2. Rumwell C. McPharlin M. Vascular Technology: An illustrated review. 4th ed. [place unknown]: Davies publishing; 2011. p. 442.
3. Katz ML, Alexandrov AV. A Practical Guide to Transcranial Doppler Examinations. [place unknown]: Summer publishing; 2003. p. 158.
4. Zemaitis MR, Boll JM, Dreyer MA. Peripheral Arterial Disease. [Updated 2023 May 23]. In: StatPearls [internet]. Treasure Island (FL): StatPearls Publishing; 2023 Jan–. Available from: https://www.ncbi.nlm.nih.gov/books/NBK430745/

11. Focused Assessment With Sonography in Trauma (FAST) Exam

11.1 Learning Objectives

After reviewing this chapter, you should be able to do the following:

1. Understand the importance of the FAST exam and the areas within the body that are assessed.
2. Explore the various techniques used in the FAST exam.

11.2 Introduction

According to the National Trauma Institute, trauma accounts for 41 million emergency department visits annually and 2.3 million hospital admissions in the United States. The Focused Assessment with Sonography in Trauma (FAST) exam is a noninvasive, quick, and accurate evaluation using bedside ultrasound to identify free fluid, usually anechoic, such as blood in the pericardial, pleural, and peritoneal spaces.[1,2] Specific indications include penetrating and blunt trauma to the chest, abdominal, and pelvic cavities. The low-frequency transducer is the most appropriate probe for the FAST exam. The patient is in a supine position. This is a simplified summary of a typical FAST examination assuming no trauma. It is imperative to first be familiar with a routine examination before examining for pathology.

11.3 Sonography for Trauma

As early as the 1990s, there was discussion in the literature regarding the usefulness of ultrasound in the rapid assessment of blunt trauma.[3] The natural advantage of ultrasound over pure clinical intervention is the accuracy of visualization, sensitivity (49–99% versus 27–45%), and specificity (95–100%). The advantage of ultrasound over CT scanning is the rapidity of the exam and the fact that the physician may remain at the bedside to make critical decisions regarding the potential rapidly changing condition of the

1. Bloom BA, Gibbons RC. Focused Assessment With Sonography for Trauma. [Updated 2023 Jul 24]. In: StatPearls [internet]. Treasure Island (FL): StatPearls Publishing; 2023 Jan–. Available from: https://www.ncbi.nlm.nih.gov/books/NBK470479/
2. Williams SR, Perera P, Gharahbaghian L. The FAST and E-FAST in 2013: Trauma ultrasonography: Overview, practical techniques, controversies, and new frontiers. Crit Care Clin. 2014 Jan;30(1):119–50, vi. doi: 10.1016/j.ccc.2013.08.005. PMID: 24295843.
3. Pearl WS, Todd KH. Ultrasonography for the initial evaluation of blunt abdominal trauma: A review of prospective trials. Ann Emerg Med. 1996 Mar;27(3):353–61. doi: 10.1016/s0196-0644(96)70273-1. PMID: 8599497.

patient. In a patient who has received blunt trauma (often from a motor vehicle accident) or penetrating trauma (such as a gunshot wound), the presence of free blood or free air in the abdomen may be seen within a few minutes by an experienced clinician.[4]

The original basis of trauma ultrasound was that the blood that escaped the vascular system due to trauma (e.g., bleeding inside the abdomen) would be seen as an abnormal hypoechoic entity in the abdomen. This concept is greatly aided by the fact that there are anatomic "potential spaces" where fluid, mostly extravascular blood, collects. Potential spaces had been known for centuries as locations where abnormal fluid collections could be reliably found and removed. It is with the more sophisticated radiology modalities that the fluid can be located before surgery and used diagnostically to help determine if surgery is necessary. Ultrasound imaging quickly and reliably indicates emergent surgery to stop the bleeding from a liver laceration, a splenic laceration, an ectopic pregnancy rupture, a bladder rupture, an injury to a major blood vessel, or less frequent injuries to other organs.

4. Boulanger BR, McLellan BA, Brenneman FD, Wherrett L, Rizoli SB, Culhane J, Hamilton P. Emergent abdominal sonography as a screening test in a new diagnostic algorithm for blunt trauma. J Trauma. 1996 Jun;40(6):867–74. doi: 10.1097/00005373-199606000-00003. PMID: 8656471.

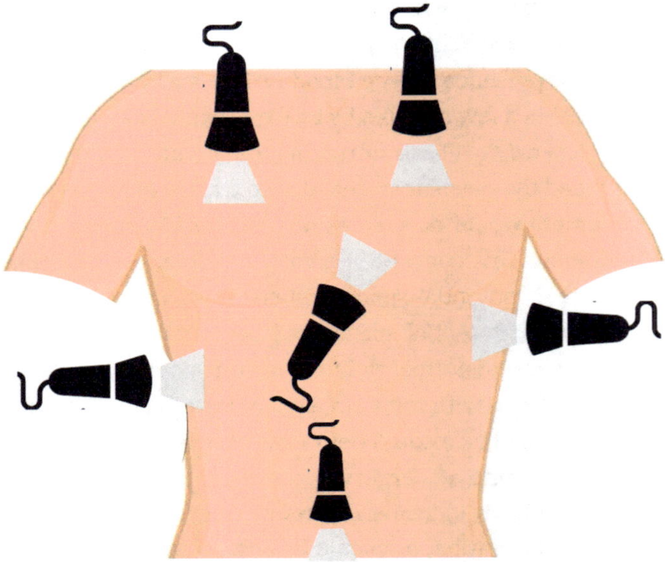

Figure 11-1: Locations that need to be examined during the FAST and eFAST exams.

The FAST exam evaluates the right upper quadrant, the left upper quadrant, and the pelvis, as shown in Figure 11-1.[5] Most famous because of the more defined areas are Morrison's pouch (in the right upper quadrant, highlighted by the contrasting bright, hyperechoic Gerota's fascia) and the pouch of Douglas (in the low midline of the pelvis between the hyperechoic borders of the bladder in males or vagina in females superiorly and the rectum inferiorly). The splenic-renal recess in the left upper quadrants and

5. Rozycki GS, Shackford SR. Ultrasound, what every trauma surgeon should know. J Trauma. 1996 Jan;40(1):1–4. doi: 10.1097/00005373-199601000-00001. PMID: 8576968.

the paracolic gutters in both the right and left lower quadrants are important to access but less easy to navigate.

The rapid ultrasound exam for trauma has extended to areas above the diaphragm as ultrasound has progressed. The eFAST (the extended FAST exam) also evaluates the lungs and heart[6] in addition to the abdomen, as shown in Figure 11-1. The probe is pointed superiorly from the xiphoid area to gain views of the heart and lungs. Again, the hypoechoic reflection of free, abnormal fluid is used as an indication of emergent surgery.

11.4 Techniques for the FAST Exam

Above the diaphragm, this fluid collection indicates emergent action in the form of a pericardial or pleural effusion. In the case of pericardial effusion, cardiac tamponade can occur and must be treated immediately. Cardiac tamponade is a medical condition that causes the restriction of ventricular filling due to pressure of fluid in the pericardium. In the case of pleural effusion, the restriction is not the initial problem. However, there may be ongoing bleeding contributing to the effusion that must be stopped. Cardiac imaging is used to evaluate pericardial effusion. There are two different cardiac views. The parasternal long-axis view is obtained by placing the transducer just left of the sternum in the fourth or fifth

6. Kirkpatrick AW, Sirois M, Laupland KB, Liu D, Rowan K, Ball CG, Hameed SM, Brown R, Simons R, Dulchavsky SA, Hamilton DR, Nicolaou S. Hand-held thoracic sonography for detecting post-traumatic pneumothoraces: The Extended Focused Assessment with Sonography for Trauma (EFAST). J Trauma. 2004 Aug;57(2):288–95. doi: 10.1097/01.ta.0000133565.88871.e4. PMID: 15345974.

intercostal space oriented to the right shoulder, as shown in Figure 11-2. The parasternal sagittal ultrasound view of the heart is shown in Figure 11-3. The abbreviations LV, RV, LA, RA, and A have been used for the left ventricle, right ventricle, left atrium, right ventricle, and aorta, respectively, in some of the following ultrasound images.

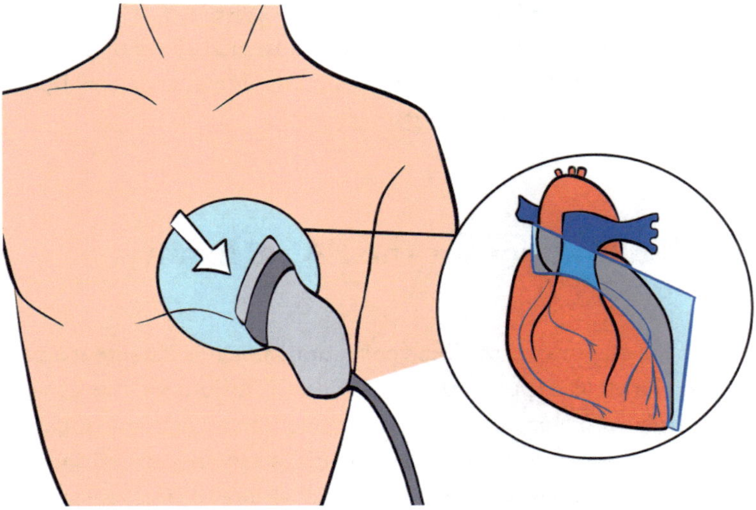

Figure 11-2: Parasternal long-axis view of the heart.

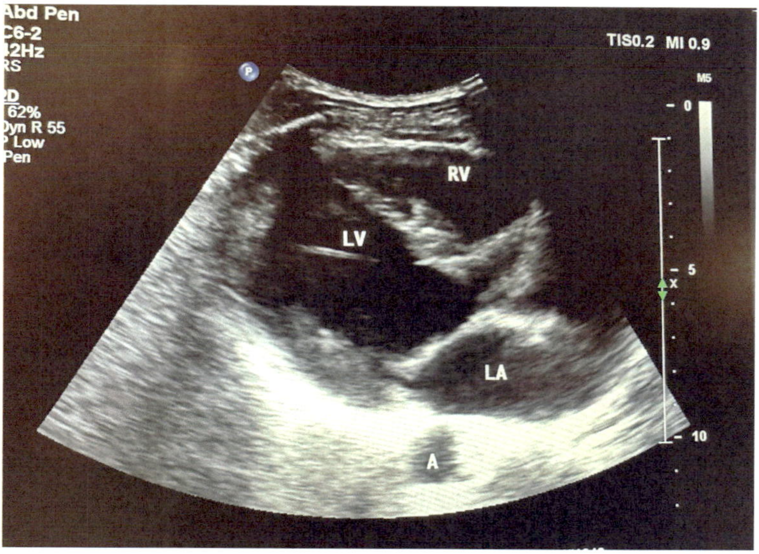

Figure 11-3: Parasternal sagittal view of the heart showing the left ventricle, right ventricle, left atrium, and aorta.

The subxiphoid view can sometimes be challenging in a patient who is considered obese. First, locate the xiphoid process, place the transducer down in a transverse position with the indicator facing toward the right, and aim between the head and left shoulder, as shown in Figure 11-4. Gently apply pressure downward, and visualize the cardiac contractility during respiration to identify the best visualization. Figure 11-5 shows the four-chamber view of the heart during this evaluation. Assess for hemopericardium, or fluid around the heart.

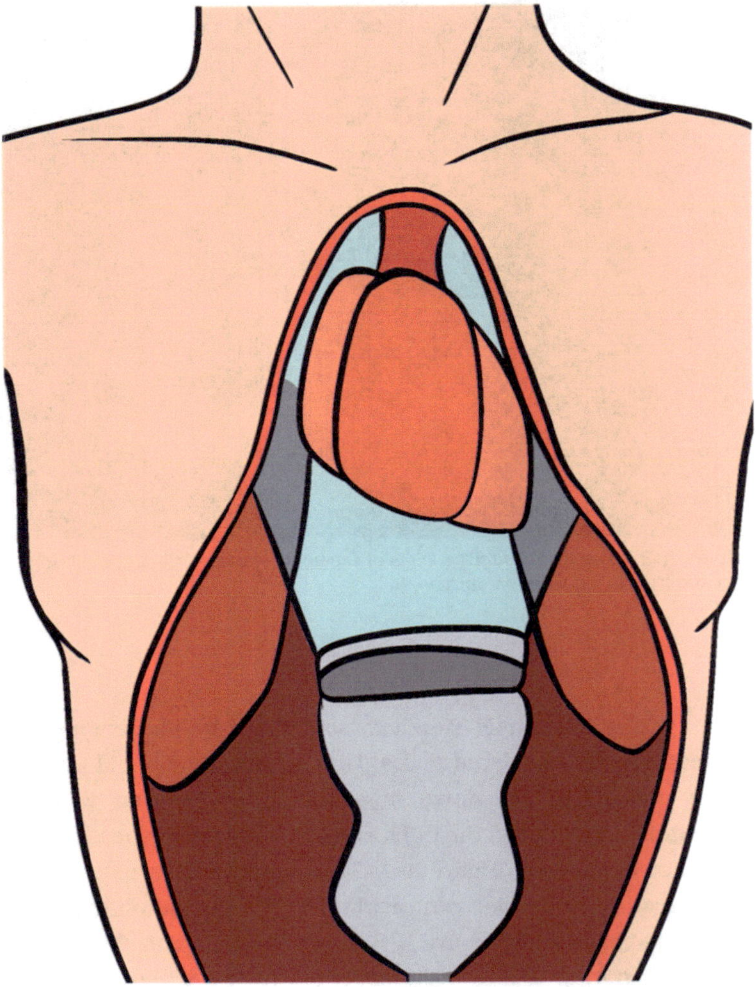

Figure 11-4: Subxiphoid view of the heart.

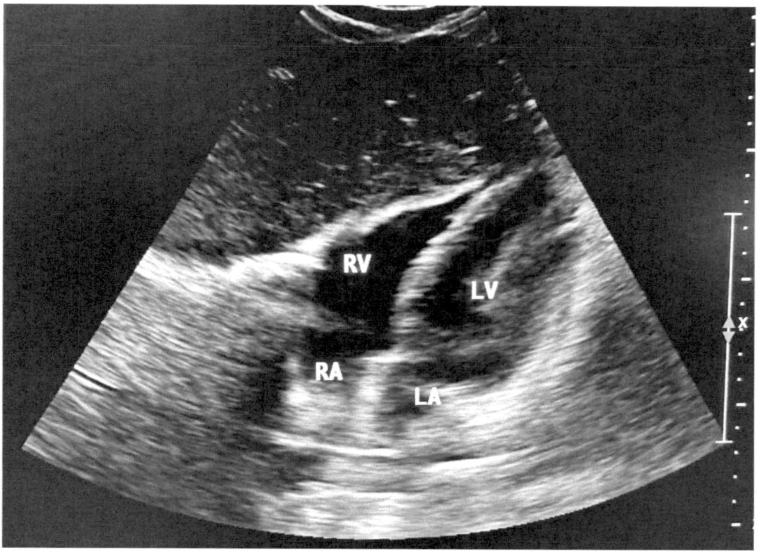

Figure 11-5: Subxiphoid four-chamber view of the heart showing the left ventricle, left atrium, right ventricle, and right atrium.

Pulmonary imaging in trauma has evolved over the years. Most recently, with the COVID-19 pandemic, it has become a more detailed and important part of the evaluation addressed in Chapter 8.

Peritoneal imaging of the abdomen/pelvis evaluates fluid in the hepatorenal recess (also referred to as Morrison's pouch), splenorenal recess, and pelvic cavity. Figure 11-6 shows a cross-sectional diagram of the abdomen, demonstrating Morrison's pouch (hepatorenal recess) and the splenorenal recess. The most straightforward abdominal view is to place the transducer in the midaxillary line at the 8th to 11th intercostal space with cephalad orientation.

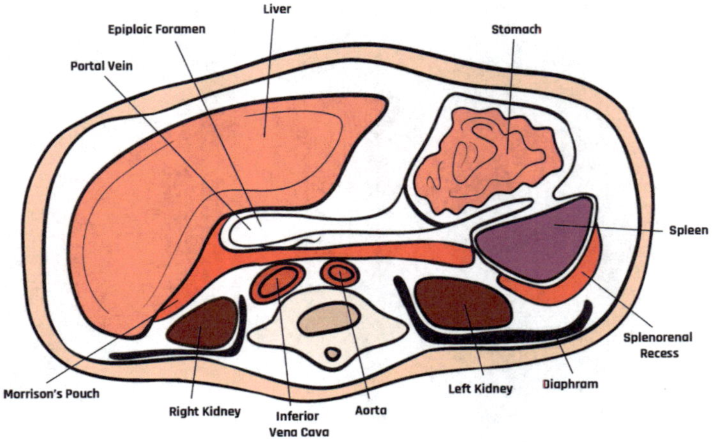

Figure 11-6: Cross-sectional diagram demonstrating Morrison's pouch (hepatorenal) and the splenorenal recess.

First, start on the right abdominal region in the midaxillary line between the 8th to 11th intercostal spaces, as shown in Figure 11-7. Maneuver the transducer by sliding, fanning, angling, and rotating until you can visualize the liver and kidney well. The abbreviations L, R, K, P, and S have been labeled in some of the following ultrasound images, representing liver, recess (hepatorenal and splenorenal), kidney, pleura, and spleen, respectively. Figure 11-8 represents the hepatorenal view, showing the liver, hepatorenal recess, kidney, and pleura.

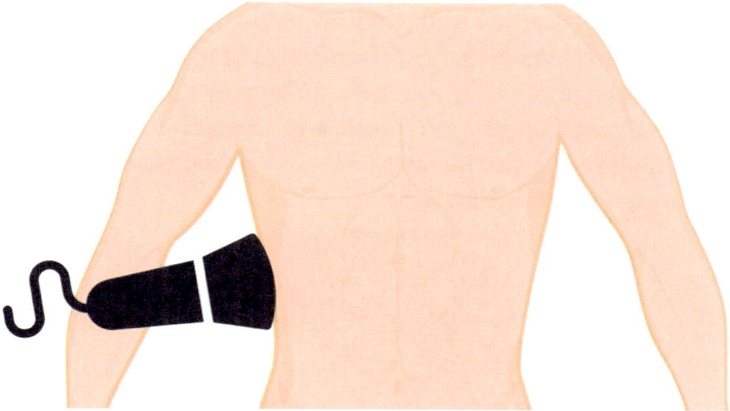

Figure 11-7: Hepatorenal view in the right upper quadrant.

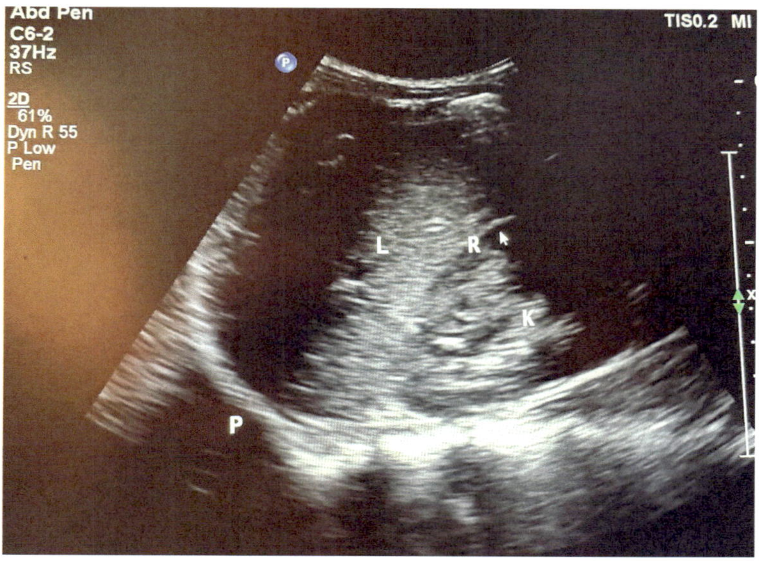

Figure 11-8: Hepatorenal view illustrating liver, hepatorenal recess (with the arrow pointing), kidney, and pleura.

Next, place the transducer in the longitudinal plane with the indicator facing the patient's head, evaluating between the fifth to ninth intercostal spaces, as shown in Figure 11-9. Again, maneuver the transducer until you can visualize the spleen and kidney well. Figure 11-10 represents the splenorenal view, showing the spleen, splenorenal recess, and kidney.

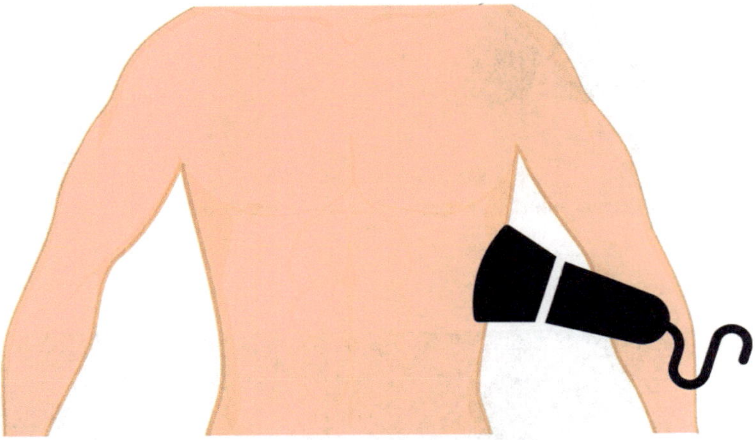

Figure 11-9: Splenorenal view in the left upper quadrant.

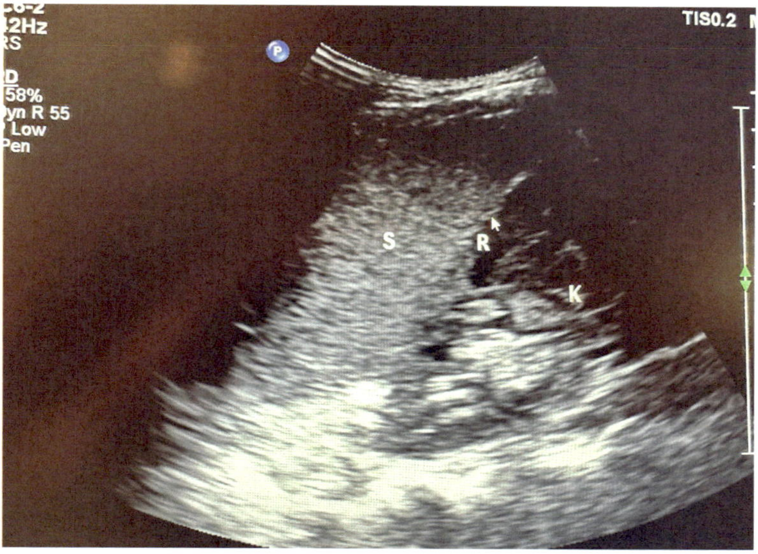

Figure 11-10: Splenorenal view illustrating spleen, splenorenal recess (with the arrow pointing), and the kidney.

The pelvic cavity also has an area where peritoneal fluid can collect within the pouch. In women, it is called the rectouterine pouch or pouch of Douglas (peritoneum between the rectum and uterus). In men, it is referred to as the rectovesical pouch (peritoneum between the rectum and bladder), as shown in Figure 11-11.

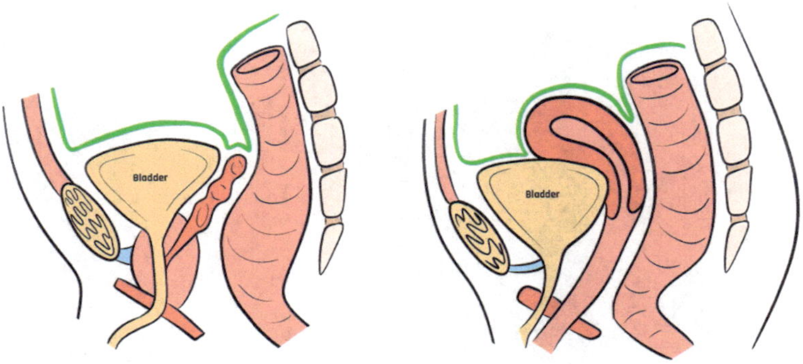

Figure 11-11: Pouches (recess) in the male and female.

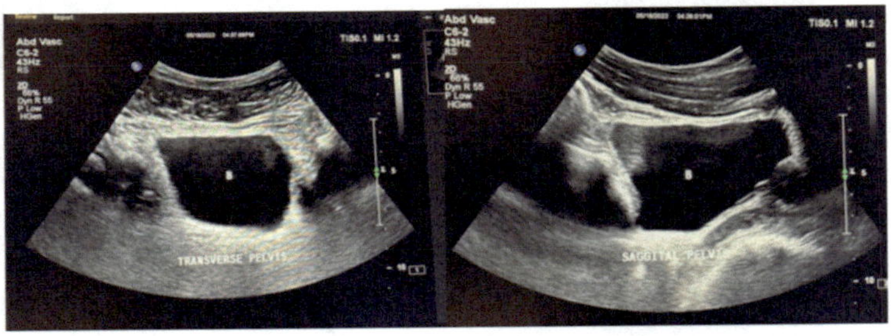

Figure 11-12: Ultrasound of the bladder.

Figure 11-12 shows an ultrasound image of the bladder (abbreviated B on the image) obtained by placing the transducer in the patient's midline, right above the pubic symphysis. This concludes the eFAST exam.

The best time to stop bleeding is as soon as possible. When the bleeding is not apparent or external, ultrasound has made leaps of progress regarding rapid diagnosis. The first hour after trauma, known as the "Golden Hour of Trauma," is a recognized time entity in which medical professionals aim to provide definitive care. In the future, it may well be a standard of care for paramedics to perform ultrasound on the scene. In this case, the image will be transmitted to an emergency department and a trauma surgeon's cell phone so that preparations for emergent surgery may be made before the patient arrives. Perhaps in the future, the patient will be taken immediately off the ambulance and directly to surgery. Trials of this technology are ongoing all over the world.

11.5 *Self-Assessment*

1. What is the original basis of the FAST or eFAST ultrasound exam for trauma?
2. At what position does one place the transducer for the parasternal long-axis cardiac view in the FAST exam?
3. How does one proceed to see the subxiphoid view in a patient who is obese, which is sometimes challenging to obtain?
4. What anatomical structures do the peritoneal imaging in the FAST exam evaluate?
5. What is another name for the hepatorenal recess?

11.6 *Further Readings*

1. Haskins SC, Bronshteyn Y, Perlas A, El-Boghdadly K, Zimmerman J, Silva M, Boretsky K, Chan V,

Kruisselbrink R, Byrne M, Hernandez N, Boublik J, Manson WC, Hogg R, Wilkinson JN, Kalagara H, Nejim J, Ramsingh D, Shankar H, Nader A, Souza D, Narouze S. American Society of Regional Anesthesia and Pain Medicine expert panel recommendations on point-of-care ultrasound education and training for regional anesthesiologists and pain physicians-part II: Recommendations. Reg Anesth Pain Med. 2021 Dec;46(12):1048–1060. doi: 10.1136/rapm-2021-102561. Epub 2021 Feb 24. PMID: 33632777.

2. Noble VN, Nelson BP. Manual of Emergency and Critical Care Ultrasound. 2nd ed. New York: Cambridge University Press; 2011. p. 360.
3. Williams SR, Perera P, Gharahbaghian L. The FAST and E-FAST in 2013: Trauma ultrasonography: Overview, practical techniques, controversies, and new frontiers. Crit Care Clin. 2014 Jan;30(1):119–50, vi. doi: 10.1016/j.ccc.2013.08.005. PMID: 24295843.
4. Griffin XL, Pullinger R. Are diagnostic peritoneal lavage or focused abdominal sonography for trauma safe screening investigations for hemodynamically stable patients after blunt abdominal trauma? A review of the literature. J Trauma. 2007 Mar;62(3):779–84.
5. Desai N, Harris T. Extended focused assessment with sonography in trauma. BJA Educ. 2018 Feb;18(2):57–62. doi: 10.1016/j.bjae.2017.

Contributors

Authors

Dr. Arbin Thapaliya
FRANKLIN COLLEGE

Dr. Arbin Thapaliya received his Ph.D. and M.S. in nuclear and particle physics from Ohio University. He also holds a M.Sc. degree in condensed matter physics and a B.Sc. in physics and mathematics from Tribhuvan University. Dr. Thapaliya has over 13 years of teaching experience in higher education. He has developed and taught courses in ultrasonography, radiation and health, and biomedical optics. Dr. Thapaliya has also authored several scientific publications in the areas of nuclear and particle physics, statistics, and diagnostic sonography.

https://orcid.org/0009-0004-9791-6230

Dr. Alec Sithole
MISSOURI WESTERN STATE UNIVERSITY

Dr. Alec Sithole received his Ph.D. in applied physics and M.S. in physics from Portland State University. He also holds an MBA (organizational management) from Eastern University and an M.Sc. in agricultural meteorology and a B.Sc. (honors) in physics from the University of Zimbabwe. Dr. Sithole has over 17 years of experience in teaching in higher education. He has authored several scientific articles and book chapters.

https://orcid.org/0000-0002-0041-3599

Dr. Michael Welsh

JOHNSON MEMORIAL HEALTH

Dr. Michael J. Welsh received his medical degree from Indiana University of the Health Sciences. He has the following certifications: family medicine, RDMS (registered diagnostic medical sonographer—abdominal and obstetrics-gynecology), RVT (registered vascular technician), and RDCS (registered diagnostic cardiac sonographer). Dr. Welsh is also the president and cofounder of Point of Care Ultrasound. He has been practicing for over 30 years in emergency and geriatric medicine. His academic interest is focused on the education of clinicians on current ultrasound modalities.

Dr. Gaston Dana

JOHNSON MEMORIAL HEALTH

Dr. Gaston Dana received his doctorate in osteopathic medicine from Nova Southeastern University of the Health Sciences. He is certified in internal medicine, emergency medicine, undersea and hyperbaric medicine, wound care and surgery, musculoskeletal and vascular sonography, and medical acupuncture. He has over 30 years of experience in practice as a physician. He uses ultrasound routinely as part of his clinical practice. He is the medical director of the Wound and Vascular Center at Johnson Memorial Hospital.

https://orcid.org/0000-0003-1820-3861

Contributor

Andrea Bearman (Project Manager)

Reviewers

Necla Erdini-Rasor
JOHNSON MEMORIAL HOSPITAL

Kyra Noerr
FRANKLIN COLLEGE

Illustrator

Katie Marshall

Made in United States
Troutdale, OR
11/03/2026

56852520R00186